De Volson Wood

The Elements of Analytical Mechanics

Second Edition

De Volson Wood

The Elements of Analytical Mechanics
Second Edition

ISBN/EAN: 9783337277482

Printed in Europe, USA, Canada, Australia, Japan

Cover: Foto ©berggeist007 / pixelio.de

More available books at **www.hansebooks.com**

THE

ELEMENTS

OF

ANALYTICAL MECHANICS.

BY

DeVOLSON WOOD, A.M., C.E.,

PROFESSOR OF MATHEMATICS AND MECHANICS IN STEVENS INSTITUTE OF TECHNOLOGY;
AUTHOR OF "RESISTANCE OF MATERIALS;" "ROOFS AND BRIDGES;" REVISED
EDITION OF "MAHAN'S CIVIL ENGINEERING," AND
"ELEMENTARY MECHANICS."

SECOND EDITION,
REVISED, CORRECTED, AND ENLARGED.

NEW YORK:
JOHN WILEY & SONS,
15 ASTOR PLACE.
1877.

PREFACE.

The plan of this edition is the same as the former one. It is designed especially for students who are beginning the study of Analytical Mechanics, and is preparatory to the higher works upon the same subject, and to Analytical Physics and Astronomy. The Calculus is freely used. I have sought to present the subject in such a manner as to familiarize the student with analytical processes. For this reason the solutions of problems have been treated as applications of general formulas. The solution by this method is often more lengthy than by special methods; still, it has advantages over the latter, because it establishes a uniformity in the process.

My experience has shown the importance of applying the fundamental equations to a great variety of problems. I have, therefore, in Article 24, and Chapters IV. and X., given a large number and a considerable variety of problems to be solved by the general equations under which they respectively fall.

In the revision I have been aided not only by my own experience with the use of the former edition in the class-room, but also by the friendly advice and criticism of several professors, that of colleges who have used the work. The result has been several pages have been rewritten, some definitions changed, and the typographical errors corrected. Several new pages in the latter part of the work have been added. I am especially indebted to Professor E. T. Quimby, of Dartmouth College, Hanover, N. H., for his valuable suggestions and for assistance in reading the final proofs.

The *nature* of force remains as much a mystery as it was

PREFACE.

when its principles were first recognized. Of its essential nature we shall probably remain forever in ignorance. We can only deal with the *laws* of its action. These laws are determined by observing the effects produced by a force. Force is the cause of an action in the physical world. The *results* of the action may be numerous and varied. Thus, force may produce pressure, tension, cohesion, adhesion, motion, affinity, polarity, electricity, etc. Or, to speak more properly, since force may be transmuted from one state to another, we would say that the above terms are names for the different manifestations of force.

The question "what is the correct measure of force" has taken different phases at different times. During the last century it was contended by some that momentum (Mv) was the correct measure, while others contended that it should be the work which it can do in a unit of time ($\tfrac{1}{2}Mv^2$). But as one has happily expressed it, "theirs was only a war of words;" for the real measure of force enters only as a factor in the expressions. Thus, if F be a constant force, the value of the momentum is Ft, see page 51, and of the work Fs, see page 45. At the present day some contend that the only measure of force is the motion which it produces, or would produce, in a unit of time. This is called the ABSOLUTE MEASURE, and THE ABSOLUTE UNIT OF FORCE *is the velocity which the force produces, or would produce, in a unit of mass in a unit of time if it acted during the unit with the intensity which it had at the instant considered.* If the intensity of the force were constant, the velocity which it produced at the end of the unit of time would be the required velocity. Hence, the absolute measure of any force acting on any mass is *the product of the mass into the acceleration;* and is the second member of equation (21). This is a correct measure, and is accepted as such by all writers on mechanics.

But those who contend that this is the only measure, necessarily deny that *weight*, or more generally *pounds*, is a measure. I contend that *pounds* is *a* measure of the intensity of a force both statically and dynamically. Many authors maintain the same position. Indeed, it is probable that the position which I have taken can be *deduced* from any standard work on mechanics; but in some it is left to inference. Thus, in Smith's *Mechanics*, page 1, we find this terse and correct definition, "The *intensity* of a force may be measured, statically, by the pressure it will produce; dynamically, by the quantity of motion it will produce." I say this is correct, but I will add that the intensity of a force which produces a given motion is also measured by a pressure, or by something equivalent to a pressure, or to a pull. To those who will look at it analytically, it is only necessary to say that the first member of equation (21) is measured in *pounds*. If we know the absolute measure, we may easily find its value in *pounds*.

The *pound* here referred to is the result of the action of gravity upon a certain quantity of matter. The amount of matter having been fixed, either by a legal enactment or by common consent, and declared to be one pound at a certain place, its weight, as determined by a standard spring-balance at any other place, becomes a measure of the force of gravity as compared with the fixed place. This standard spring-balance may measure the intensity in pounds of any other force, whether the body upon which the force acts be at rest or in motion. If a perfectly free body were placed in a hollow space at the centre of the earth, at which place it would be devoid of weight, and pulled or pushed by a constant force, whose intensity, measured by a standard spring-balance, equaled the weight of an equal body on the surface of the earth, then would its motion be the same as that of a falling body. See page 24, Problem 7. In the forces of nature producing motion, there being no visible connection between the point of action of the force and the body upon which it acts, we are unable to *weigh* their intensity except by calculation. If the absolute measure is known, the *pounds* of intensity may be computed. The absolute measure of the force of gravity on a mass m is mg, and the weight of the body being W, we have $W = mg$. The sun acts upon the earth with a force which may be expressed by the absolute measure, and also by a certain number of pounds of force. More than half of the examples in Article 24 involve an equality between *pounds of intensity* and the absolute measure of the force. The fact is, that, in case of motion, these quantities are co-relative. Since, then, it is correct to use the term *pound* as the measure of the intensity of a force whether the body be at rest or in motion, and since it is in common use, and the student is familiar with it, I prefer to consider a force as measured by a certain number of pounds. See Article 9. It is more simple, containing as it does only one element, than the absolute measure, which contains three elements—mass, velocity, and time.

There is another advantage in thus measuring force. Students frequently, and in some cases writers, use the expressions, "quantity of force," "amount of force," "force of a blow," etc., when they mean (or should mean) momentum, or work, or vis viva. In such cases an attempt to answer the question "how many pounds of force" would show at once that the quantity referred to was not *force*.

So much ambiguity, or at least indefiniteness, has arisen in regard to the term force, that I have rejected the terms "Impulsive Force" and "Instantaneous Force," and used the term "Impulse" instead of them. We know nothing of an instantaneous force, that is, one which requires no time for its action. I also reject the expression *force of inertia*. I do not believe that *inertia* is a *force*. To the question "The inertia of a body is how many pounds of force" there is no answer.

The term *moment of inertia* has no physical representation. The nearest approach to it is in the expression for the vis viva of a rotating body. In such problems the moment of inertia forms an important factor. The energy of a rotating body having a constant angular velocity is directly proportional to its moment of inertia in reference to its axis of rotation. See page 202. But *motion* is not necessary for its existance. See page 165. The expression appears in the discussion of numerous statical problems, such as the flexure of a beam, the centre of pressure of a fluid, the centre of gravity of certain solids, etc. It is not the moment of a moment, although it may be so construed as to appear to be of that *form*. Some other term might be more appropriate. Even the expression *moment of the mass* would be less objectionable.

The subjects of *Centrifugal Force* and *Unbalanced Force* have been discussed of late in *Engineering*. Some assert that there is no such thing as a centrifugal force. Much unprofitable discussion may be avoided by strictly defining the terms used. If it is defined to be a force equal and opposite to the deflecting force, it will, at least, have an ideal existence, just as the resultant in statical problems has an ideal existence. But the vital question is, is the centrifugal force active when the deflecting force acts? Or, in other words, do both act upon the body at the same time? It seems, however, quite evident that if both acted upon the body at the same time they would neutralize each other, and the body would move in a straight line. Hence, in the movement of the planets, or of any free rotating body, there is no centrifugal force. But in the case of a locomotive running around a curve there may be both centripetal and centrifugal forces; the former acting against the locomotive to force it away from a tangent to the track; the latter, against the track, tending to force it outward. Wherever the force is conceived to act, whether just between the rail and wheel or at some other point, it is evident that both do not act upon the same body.

Similarly in regard to the *unbalanced force*. It is a convenient term to use, but, in a strict sense, an unbalanced force does not exist; for action and reaction are equal and opposite. But in reference to a particular body, all other conditions being ignored, the force may be unbalanced. Thus, when a ball is fired from a cannon, the force of the powder, considered in the direction of the motion of the ball only, is unbalanced; but the powder exerts an equal force in the opposite direction, and in that sense also is unbalanced. But when the entire effect of the force in all directions is considered, the algebraic resultant is zero. In other words, the centre of gravity of the system, for forces acting between its integrant parts, remains constant.

These are some of the fundamental questions which will arise in the mind of the student as he studies the subject. Fortunately, it is not necessary for him to settle them beyond the question of a doubt before he proceeds with the subject. On some of these points scholars, who have made the subject a specialty, differ; and it is only after a careful consideration of the points involved that one can take an intelligent position in regard to them.

DeVolson Wood.

Hoboken, *August*, 1877.

TABLE OF CONTENTS.

CHAPTER I.
DEFINITIONS AND THE LAWS OF FORCES WHICH ACT ALONG A STRAIGHT LINE.

ARTICLES	PAGE
1–14—Definitions	1
15–20—Velocity; Gravity; Mass	6
21–23—Force; Mass; Density	16
24—Examples of Accelerated Motion	21
25–27—Work; Energy; Momentum	44
28–33—Impact	52
34–36—Statics; Power; Inertia	58
37–38—Newton's Laws of Motion. Eccentric Impact	62

CHAPTER II.
COMPOSITION AND RESOLUTION OF FORCES.

39–42—Conspiring Forces	65
43–46—Composition of Forces	65
47—Resolution on Three Axes	70
48–49—Constrained Equilibrium of a Particle	71
50–60—Statical Moments	79
Examples	84

CHAPTER III.
PARALLEL FORCES.

61–63—Resultant	86
64—Moments of Parallel Forces	87
65–68—Statical Couples	88
69–74—Centre of Gravity of Bodies	92
75—Centrobaric Method	103
77–81—General Properties of the Centre of Gravity	108
82—Centre of Mass	110

TABLE OF CONTENTS.

CHAPTER IV.
NONCONCURRENT FORCES.

ARTICLES PAGE
83- 85—General Equations.. 111
86—Problems 115

CHAPTER V.
STRESSES.

87- 90—Stresses Resolved... 142
91- 92—Shearing Stresses—Notation................................... 144
93- 94—Resultant Stress... 146
95- 96—Discussion 158
97- 98—Conjugate Stresses. General Problem..................... 156

CHAPTER VI.
VIRTUAL VELOCITIES.

100—Concurring Forces... 159
101—Nonconcurring Forces.. 160
    Examples.. 161

CHAPTER VII.
MOMENT OF INERTIA.

102-104—Examples. Formula of Reduction......................... 165
105—Relation between Rectangular Moments..................... 169
106—Moments of Inertia of Solids...................................... 172
107—Radius of Gyration.. 173

CHAPTER VIII.
MOTION OF A FREE PARTICLE.

108—General Equations... 175
109—Velocity and Living Force... 176
110-117—Central Forces.. 180

CHAPTER IX.
CONSTRAINED MOTION OF A PARTICLE.

118-121—General Equations... 191
122-124—Centrifugal Force on the Earth............................ 198

CHAPTER X.

ROTATION OF A BODY WHEN THE FORCES ARE IN A PLANE.

ARTICLES	PAGE
125–131—General Equations	200
132—Reduced Mass	204
133–135—Spontaneous Axis; Centre of Percussion; Axis of Instantaneous Rotation	209
Examples	210
136–138—Compound Pendulum. Captain Kater's Experiments	218
139–142—Ellipticity of the Earth. Torsion Pendulum	221
143—Density of the Earth	224
144–146—Problems	227

CHAPTER XI.

GENERAL EQUATIONS OF MOTION.

147—D'Alembert's Principle	230
148—General Equations	231
149—Conservation of the Centre of Gravity	235
150—Conservation of Areas	235
151—Conservation of Energy	236
152—Composition of Angular Velocities	240
153—Moments of Rotation of the Centre	242
154—Motion of the Centre of a Body	242
155—Motion of Rotation of the Centre	243
156—Euler's Equations	243
157—Principal Axes	244
158—No Strain on Principal Axes	246
159—Relation between the Fixed and Principal Axes	247
160—Relations between Angular Velocities	247

GREEK ALPHABET.

Letters.	Names.	Letters.	Names.
A α	Alpha	N ν	Nu
B β	Bēta	Ξ ξ	Xi
Γ γ	Gamma	O o	Omicron
Δ δ	Delta	Π π	Pi
E ε	Epsilon	P ρ	Rho
Z ζ	Zēta	Σ σ ς	Sigma
H η	Eta	T τ	Tau
Θ ϑ θ	Thĕta	Υ υ	Upsilon
I ι	Iōta	Φ φ	Phi
K κ	Kappa	X χ	Chi
Λ λ	Lambda	Ψ ψ	Psi
M μ	Mu	Ω ω	Omega

ANALYTICAL MECHANICS.

CHAPTER I.

DEFINITIONS, AND PRINCIPLES OF ACTION OF A SINGLE FORCE, AND OF FORCES ACTING ALONG THE SAME LINE.

1. MECHANICS treats of the laws of forces, and the equilibrium and motion of bodies under the action of forces. It has two grand divisions, Dynamics and Statics.

2. DYNAMICS treats of the motion of material bodies, and the laws of the forces which govern their motion.

3. STATICS treats of the conditions of the equilibrium of bodies under the action of forces.

There are many subdivisions of the subject, such as Hydrodynamics, Hydrostatics, Pneumatics, Thermodynamics, Molecular Mechanics, etc. That part of mechanics which treats of the relative motion of bodies which are so connected that one drives the other, such as wheels, pulleys, links, etc., in machinery, is called Cinematics. The motion in this case is independent of the intensity of the force which produces the motion.

Theoretic Mechanics treats of the effect of forces applied to material points or particles regarded as without weight or magnitude. Somatology is the application of theoretic mechanics to bodies of definite form and magnitude.

4. MATTER is that which receives and transmits force. In a physical sense it possesses extension, divisibility, and impenetrability.

Matter is not confined to the gross materials which we see and handle, but includes those substances by which sound, heat, light, and electricity are transmitted.

It is unnecessary in this connection to consider those refined speculations by which it is sought to determine the essential nature of matter. According to some of these speculations, *matter* does not exist, but is only a conception.

According to this view, bodies are forces, within the limit of which the attractive exceed the repulsive ones, and at the limits of which they are equal to each other.

But observation, long continued, teaches practically that matter is inert, that it has no power within itself to change its condition in regard to rest or motion; that when in motion it cannot change its rate of motion, nor be brought to rest without an external cause, and this cause we call FORCE. One also learns from observation that matter will transmit a force, as for instance a pull applied at one end of a bar or rope is transmitted to the other end; also a moving body carries the *effect* of a force from one place to another.

5. A BODY is a definite portion of matter. A particle is an infinitesimal portion of a body, and is treated geometrically as a point. A molecule is composed of several particles. An atom is an *indivisible* particle.

6. FORCE is that which tends to change the state of a body in regard to rest or motion. It moves or tends to move a body, or change its rate of motion.

We know nothing of the *essential* nature of force. We deal only with the *laws* of its action. These laws are deduced by observations upon the effects of forces, and on the hypothesis that *action and reaction are equal and opposite;* or, in other words, that the effect equals the cause. In this way we find that forces have different intensities, and that a relation may be established between them. It is necessary, therefore, to establish a UNIT. This may be assumed as the effect of any known force, or a multiple part thereof. The effect of all known forces is to produce a pull, or push, or their equivalents, and may be measured by pounds, or by something equivalent. The force of gravity causes the weight of bodies, and this is measured by pounds. We therefore assume that a STANDARD POUND is *the* UNIT of force.

The standard pound is established by a legal enactment, and has been so fixed that a cubic foot of distilled water at the level of the sea, at latitude 45 degrees, at a temperature of 62 degrees Fahrenheit, with the barometer at 30 inches, will weigh about 62.4 pounds avoirdupois.

The English standard pound was originally 5,760 troy grains. The grain was the weight of a certain piece of brass which was deposited with the clerk of the House of Commons. This was destroyed at the time of the burning of the House of Commons in 1834, after which it was decided that the legal pound should be the weight of a certain piece of platinum, weighing 7,000 grains. This is known as the avoirdupois pound, and the troy pound ceased to be the legal standard, although both have remained in common use.

The legal standard pound in the United States is a copy of the English troy pound, and was deposited in the United States Mint in Philadelphia, in 1827, where it has remained. The avoirdupois pound, or 7,000 grains, is used in nearly all commercial transactions. The troy pound is a standard at 62 degrees Fahrenheit and 30 inches of the barometer.

The weight of a cubic inch of water at its maximum density, as accepted by the Bureau of Weights and Measures of the United States, is stated by Mr. Hasler, in a report to the Secretary of the Treasury, 1842, to be 252.7453 grains. Mr. Hasler determined the temperature at which water has a maximum density, at 39.83 degrees Fahrenheit, but Playfair and Joule determined it to be 39.101° F.

The *exact* determination of the equivalent values of the units is very difficult, and has been the subject of much scientific investigation.—(See *The Metric System*, by F. A. P. Barnard, LL.D., New York, 1872.)

When a quantity can be measured directly, the *unit* is generally of the same quality as the thing to be measured : thus, the unit of time is time, as a day or second; the *unit* of length is length, as one inch, foot, yard, or metre; the *unit* of volume is volume, as one cubic foot; the *unit* of money is money; of weight is weight; of momentum is momentum; of work is work, etc.

When dissimilar quantities are used to measure each other a proportion must be established between them. It is commonly said that "the arc measures the angle at the centre," but it does not do it directly, since there is no ratio between them. The arc is a linear quantity, as feet or yards, or a number of times the radius, while the angle is the divergence of two lines, and is usually expressed in degrees. But angles are *proportional* to their subtended arcs; hence we have an equality of ratios, or

$$\frac{angle}{unit\ angle} = \frac{subtended\ arc}{arc\ which\ subtends\ the\ unit\ angle};$$

and since a semi-circumference, or π, subtends an angle of 180°, it is easy from the above equality of ratios to determine any angle when the arc is known, or *vice versâ*.

Similarly, the intensity of heat is not measured directly, but by its effect in expanding liquids or metals.

The magnetic force is measured by its effect upon a magnetic needle.

The intensities of lights by the relative shadows produced by them.

Similarly with forces, we measure them by their effects.

Dissimilar quantities, between which no proportion exists, do not measure each other. Thus feet do not measure time, nor money weight.

Pounds for commercial purposes represents quantities of matter; but when applied to forces it represents their intensities. In a strict sense, *pounds* does not measure *directly* the quantity of matter, but is always a measure of a force.

7. THE LINE OF ACTION of a force is the line along which the force moves or tends to move a particle. If the particle is acted upon by a single force, the line of action is straight. This is also called the *action-line* of the force.

8. THE POINT OF APPLICATION of a force is the point at which it acts. This may be considered as at any point of its action-line. Thus, if a pull be applied at one end of a cord, the effect at the other end is the same as if applied at any intermediate point.

9. A FORCE is said to be given when the following elements are known:—

1st. Its magnitude (*pounds*);
2d. The direction of the line along which it acts (*action-line*);
3d. The direction along the action-line (+ or —); and,
4th. Its point of application.

A force may be definitely represented by a straight line; thus, its magnitude may be represented by the length AB, Fig. 1; its position by the position of the line AB; its direction along the line by the arrow-head at B, which indicates that the force acts from A towards B; and its point of application by the end A.

FIG. 1.

10. SPACE is indefinite extension, finite portions of which may be measured.

11. TIME is duration, and may be measured.

Probably no definition will give a better idea of the abstract quantities of *time* and *space* than that which is formed from experience.

12. A BODY is in motion when it occupies successive portions of space in successive instants of time. In all other cases it is at *absolute* rest. Motion in reference to another moving body is *relative*.

But a body may be at rest in reference to surrounding objects and yet be in motion. Thus, many objects on the surface of the earth, such as rocks, trees, etc., may be at rest in reference to objects around them, while they move with the earth through space. Observation teaches that there is probably no body at absolute rest in the universe.

13. MOTION IS UNIFORM when the body passes over equal portions of space in equal successive portions of time.

14. VARIABLE MOTION is that in which the body passes over unequal portions of space in equal times.

15. Velocity *is the rate of motion.* When the motion is *uniform* it is measured by the linear distance over which a body would pass in a unit of time; and when it is *variable* it is the distance over which it would pass if it moved with the rate which it had at the instant considered. The path of a moving particle is the line which it generates.

For uniform velocity, we have

$$v = \frac{s}{t}, \qquad (1)$$

in which

$s =$ the space passed over;
$t =$ the time occupied in moving over the space s; and
$v =$ the velocity.

For variable velocity, we have

$$v = \frac{ds}{dt}. \qquad (2)$$

Examples.

1. If a particle moves uniformly thirty feet in three seconds, what is its velocity?

2. If $s = at$, what is the velocity?

3. If $s = at^2 + bt$, what is the velocity at the time t, or at the end of the space s?

Here

$$v = \frac{ds}{dt} = 2at + b,$$

which is the answer to the first part. Find t from the given equation, and substitute in the expression for v, and it gives the answer to the second part; or

$$v = \sqrt{b^2 + 4as}.$$

4. If $s = 4t^3$, required the velocity at the end of five seconds.

5. If $3s^3 = 5t^2$, required the velocity at the end of ten seconds.

6. If $s = \frac{1}{2}gt^2$, what is the velocity in terms of the time and space?

7. If $at = e^{bs} - 1$, required the velocity in terms of the time and space.

16. ANGULAR VELOCITY *is the rate of angular movement.* If a particle moves around a point having either a constant or a variable velocity along its path, the angular velocity is measured by the arc at a unit's distance which subtends the angle swept over in a unit of time by that radius vector which passes through the particle.

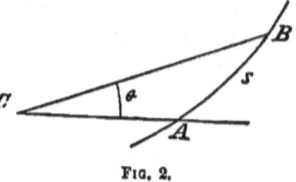

Fig. 2.

If $s = AB =$ the length of the path described;
$v =$ the velocity along the path AB;
$t =$ the time of the movement;
$r = CB =$ the radius vector;
$\theta =$ the circular arc at a unit's distance which subtends the angle ACB swept over by the radius vector in the time t; and
$\omega =$ the angular velocity.

Then, if the angular motion is *uniform*,

$$\omega = \frac{\theta}{t}. \qquad (3)$$

If it is *variable*, then

$$\omega = \frac{d\theta}{dt}. \qquad (4)$$

We also have,

$$ds = vdt = \sqrt{r^2 d\theta^2 + dr^2};$$

$$\therefore d\theta^2 = \frac{ds^2 - dr^2}{r^2} = \frac{v^2 dt^2 - dr^2}{r^2}; \text{ and}$$

$$\omega = \frac{d\theta}{dt} = \frac{\sqrt{v^2 - \left(\frac{dr}{dt}\right)^2}}{r}. \qquad (5)$$

17. ACCELERATION *is the rate of increase or decrease of the velocity.* It is a *velocity-increment*. The *velocity-increment* of an increasing velocity is considered positive, and that of a decreasing velocity, negative.

The measure of the acceleration, when it is uniform, is the amount by which the velocity is increased (or decreased) in a unit of time. If the acceleration is variable, it is the amount by which the velocity would be increased in a unit of time, provided the rate of increase continued the same that it was at the instant considered.

Hence, if
$$f = \text{the measure of the acceleration (or, briefly, the acceleration)};$$
then, when the acceleration is uniform,
$$f = \frac{v}{t} = \frac{\frac{ds}{dt}}{t},$$
and hence, when it is variable,
$$f = \frac{dv}{dt} = \frac{d\frac{ds}{dt}}{dt} = \frac{d^2s}{dt^2}. \tag{6}$$

We also have
$$f = \frac{d^2s}{dt^2} \times \frac{ds^2}{ds^2} = \frac{d^2s}{ds^2} \times \frac{ds^2}{dt^2} = v^2 \frac{d^2s}{ds^2}. \tag{6'}$$

We thus see that the relation between space, time, and velocity are independent of the cause which produces the velocity.

APPLICATIONS OF EQUATION (6).

1. Suppose that the acceleration is constant.

Then in (6) f will be constant, and dt being the equicrescent variable, we have
$$f \int dt = \frac{1}{dt} \int d^2s,$$
$$\text{or } ft = \frac{ds}{dt} + C_1. \tag{7}$$

But for $t = 0$, $\frac{ds}{dt} = v_0 =$ the initial velocity $\therefore C_1 = -v_0$; and (7) becomes
$$ds = ftdt + v_0 dt.$$

Integrating again gives
$$s = \tfrac{1}{2}ft^2 + v_0 t + C_2.$$

But $s = s_0$ for $t = 0$ ∴ $C_2 = s_0$;
hence the final equation is
$$s = \tfrac{1}{2}ft^2 + v_0 t + s_0; \qquad (8)$$
which gives the relation between the space and time.

Again, multiplying both members of equation (6) by ds, we have
$$\frac{1}{dt^2}\int ds\, d^2s = f\int ds\,;$$

or
$$\frac{ds^2}{dt^2} = 2fs + v_0^2; \qquad (9)$$

$$\therefore dt = \frac{ds}{\sqrt{v_0^2 + 2fs}};$$

and integrating, gives
$$t = \frac{\sqrt{v_0^2 + 2fs}}{f} + C. \qquad (10)$$

Equation (7) gives the relation between the velocity and time, and equation (9) between the velocity and space.

If v_0 and s_0 are both zero, the preceding equations become

$$v = ft = \sqrt{2fs} = \frac{2s}{t}. \qquad (11)$$

$$s = \tfrac{1}{2}ft^2 = \frac{v^2}{2f} = \tfrac{1}{2}vt. \qquad (12)$$

$$t = \frac{v}{f} = \sqrt{\frac{2s}{f}} = \frac{2s}{v}. \qquad (13)$$

We shall find hereafter that these formulas are applicable to all cases in which the *force* is constant and uniform.

2. Find the relation between the space and time when the acceleration is naught.

We have
$$\frac{d^2s}{dt^2} = 0.$$

Multiply by dt, integrate twice, and we have
$$s = s_0 + v_0 t;$$
in which s_0 and v_0 are initial values; that is, the body will have

passed over a space s_0 before t is computed. v_0 is not only the initial but the constant uniform velocity. If $s_0 = 0$, then $s = v_0 t$.

3. If the acceleration varies directly as the time from a state of rest, required the velocity and space at the end of the time t.

Here $f = at$.

4. Determine the velocity when the acceleration varies inversely as the distance from the origin and is negative; or $f = -\dfrac{a}{s}$.

5. Determine the relation between the space and time when the acceleration is negative and varies directly as the distance from the origin; or $f = -bs$.

Equation (6) becomes

$$\frac{d^2s}{dt^2} = -bs.$$

Multiplying both members by ds, we have

$$\frac{ds\, d^2s}{dt^2} = -bs\, ds.$$

Integrating gives

$$v^2 = \frac{ds^2}{dt^2} = -bs^2 + C_1.$$

But $v = 0$ for $s = s_0$ ∴ $C_1 = bs_0^2$; and

$$\frac{ds^2}{dt^2} = b(s_0^2 - s^2) = v^2, \qquad (14)$$

or $b^{\frac{1}{2}} dt = \dfrac{ds}{(s_0^2 - s^2)^{\frac{1}{2}}}.$

Integrating again gives

$$b^{\frac{1}{2}} t = \sin^{-1} \frac{s}{s_0} + C_2.$$

But $t = 0$ for $s = s_0$ ∴ $C_2 = -\tfrac{1}{2}\pi$;

$$\therefore s = s_0 \sin(tb^{\frac{1}{2}} + \tfrac{1}{2}\pi). \qquad (15)$$

If $s = s_0, t = 0,$ $\quad 2\pi b^{-\frac{1}{2}},$ $\quad 4\pi b^{-\frac{1}{2}}.$

" $s = 0, t = \frac{1}{2}\pi b^{-\frac{1}{2}}, \frac{3}{2}\pi b^{-\frac{1}{2}}, \frac{5}{2}\pi b^{-\frac{1}{2}}, \frac{7}{2}\pi b^{-\frac{1}{2}}, \frac{9}{2}\pi b^{-\frac{1}{2}},$ etc.;

" $s = s_0, t = \quad\quad \pi b^{-\frac{1}{2}},\quad\quad 3\pi b^{-\frac{1}{2}},\quad\quad 5\pi b^{-\frac{1}{2}}.$

This is an example of periodic motion, of which we shall have examples hereafter.

6. Determine the space when the acceleration diminishes as the square of the velocity.

When the acceleration is constant, the relation between the time, space, and velocity may be shown by a triangle, as in Fig. 3. Let AB represent the time, say four seconds. Divide it into four equal spaces, and each space will represent a second. Draw horizontal lines through the points of division and limit them by the inclined line AC. The horizontal lines will represent the corresponding velocities. Thus $v_2 = ge$ is the velocity at the end of the time t_2. The triangle Abc represents the space passed over during the first second, and ABC the space passed over during four seconds. The lines de, fh, and iC represent the accelerations for each second, which in this case are equal to each other, and equal to bc, which is the velocity at the end of the first second. Hence, *when the acceleration is uniform, the velocity at the end of the first second equals the acceleration.* This is also shown by Eq. (11); for if $t = 1, v = f$. Equations (12) and (13) may be deduced directly from the figure.

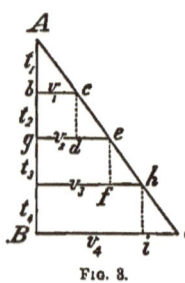
Fig. 3.

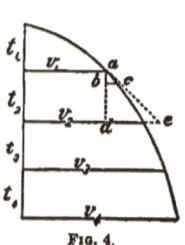
Fig. 4.

If acceleration constantly varies, the case may be represented as in Fig. 4. To find the acceleration at the end of the first second, draw a tangent ae to the curve at the point a, and drop the perpendicular ad, then will de be the acceleration. But $\dfrac{de}{ad} = \dfrac{bc}{ab} = \dfrac{dv}{dt} = f =$ the velocity-increment, which is the same as Equation (6).

18. Resolved Velocities and Accelerations. When the motion is along a known path and at a known rate, the projections of the velocities and accelerations upon other paths which are inclined to the given one will equal the product of these quantities by the cosine of the angle between the paths; that is,

$$v' = v \cos \theta = \frac{ds}{dt} \cos \theta, \text{ and } f' = \frac{d^2s}{dt^2} \cos \theta,$$

where v' and f' are on the new path, and θ the angle between the paths.

Examples.

1. If the velocity v is constant and along the line AB, which makes an angle θ with the line AC, then will the velocity projected on AC also be constant, and equal to $v \cos \theta$; and on the line BC, equal to $v \sin \theta$.

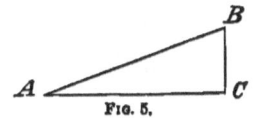

Fig. 5.

2. Let ABC be a parabola whose equation is $y^2 = 2px$. If a body describes the arc BC with such a varying velocity that its projection on BD, a tangent at B, is constant, required the velocity and the acceleration parallel to BE.

From the equation of the curve we have

$$\frac{dy}{dx} = \frac{p}{y}.$$

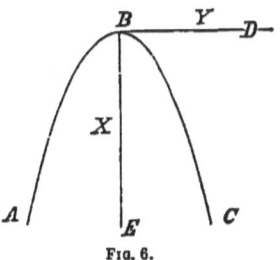

Fig. 6.

From the conditions of the problem we have

$$\frac{dy}{dt} = \text{constant} = v';$$

but

$$\frac{dx}{dt} = \frac{dy}{dt} \frac{dx}{dy} = v' \frac{y}{p} = v'\sqrt{\frac{2x}{p}},$$

which is the velocity parallel to x;

$$\therefore \frac{d^2x}{dt^2} = \frac{v'}{p}\frac{dy}{dt} = \frac{v'^2}{p};$$

hence the acceleration parallel to the axis of x will be constant.

Let ds be an element of the arc, then will the velocity along the arc be

$$v = \frac{ds}{dt} = \left(\frac{dx^2}{dt^2} + \frac{dy^2}{dt^2}\right)^{\frac{1}{2}} = v'\left(1 + \frac{2x}{p}\right)^{\frac{1}{2}}.$$

3. Determine the accelerations parallel to the co-ordinate axes x and y, so that a particle may describe the arc of a parabola with a constant velocity.

Let the equation of the parabola be

$$y^2 = 2px;$$

$$\therefore \frac{dy}{dx} = \frac{p}{y}.$$

The conditions of the problem give

$$\frac{ds}{dt} = \text{constant} = v.$$

But, $\dfrac{ds}{dt} = \dfrac{dy}{dt} \cdot \dfrac{ds}{dy} = \dfrac{dy}{dt}\dfrac{\sqrt{dx^2 + dy^2}}{dy} = \dfrac{dy}{dt}\sqrt{1 + \dfrac{dx^2}{dy^2}}$

$$= \frac{dy}{dt}\sqrt{1 + \frac{y^2}{p^2}} = v;$$

$$\therefore \frac{dy}{dt} = \frac{pv}{\sqrt{p^2 + y^2}};$$

and differentiating, gives

$$\frac{d^2y}{dt^2} = -\frac{pvy}{(p^2 + y^2)^{\frac{3}{2}}}\frac{dy}{dt}$$

$$= -\frac{p^2v^2y}{(p^2 + y^2)^2};$$

which being negative shows that the acceleration perpendicular to the axis of the parabola constantly diminishes.

Similarly we find

$$\frac{d^2x}{dt^2} = \frac{p^3v^2}{(y^2 + p^2)^2}.$$

EXAMPLES.

4. A wheel rolls along a straight line with a uniform velocity; compare the velocity of any point in the circumference with that of the centre.

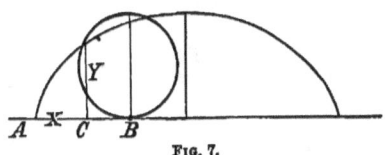

Fig. 7.

Let $v =$ the velocity of any point in the circumference,
$v' =$ the uniform velocity of the centre,
$r =$ the radius of the circle,
$x =$ the abscissa which coincides with the line on which it rolls, and
$y =$ the ordinate to any point of the cycloid.

Take the origin at A. The centre of the circle moves at the same rate as the successive points of contact B. The centre is vertically over B. The abscissa of the point of contact corresponding to any ordinate y of the cycloid, is $r\ versin^{-1}\frac{y}{r}$;

$$\therefore v' = \frac{d}{dt}\left(r\ versin^{-1}\frac{y}{r}\right) = \frac{r}{\sqrt{2ry - y^2}} \cdot \frac{dy}{dt};$$

$$\therefore \frac{dy}{dt} = v'\frac{\sqrt{2ry - y^2}}{r}.$$

The equation of the cycloid is

$$x = r\ versin^{-1}\frac{y}{r} - (2ry - y^2)^{\frac{1}{2}};$$

and from the theory of curves

$$ds^2 = dx^2 + dy^2 \therefore \frac{ds}{dy} = \sqrt{1 + \frac{dx^2}{dy^2}};$$

or,
$$\frac{ds}{dy} = \sqrt{\frac{2r}{2r - y}};$$

and,
$$v = \frac{ds}{dt} = \frac{dy}{dt} \cdot \frac{ds}{dy} = \sqrt{\frac{2y}{r}} \cdot v'.$$

If
$$y = 0, \quad v = 0;$$
$$y = r, \quad v = \sqrt{2}v';$$
$$y = 2r, \quad v = 2v';$$
$$y = \tfrac{1}{2}r, \quad v = v'.$$

Hence, at the instant that any point of the wheel is in contact with the straight line, it has no velocity, and the velocity at the highest point is twice that of the centre.

The velocity at any point of the cycloid is the same as if the wheel revolved about the point of contact, and with the same angular velocity as that of the generating circle.

For, the length of the chord which corresponds to the ordinate y is $\sqrt{2ry}$, and hence, if

$$v : 2v' :: \sqrt{2ry} : 2r ;$$

we have $\quad v = \sqrt{\dfrac{2y}{r}}\, v'$, as before found.

19. Gravitation is that natural force which mutually draws two bodies towards each other. It is supposed to extend to every particle throughout the universe according to fixed laws. The force of gravity above the surface of the earth diminishes as the square of the distance from the centre increases, but within the surface it varies directly as the distance from the centre. If a body were elevated one mile above the surface of the earth it would lose nearly $\frac{1}{2000}$ of its weight, which is so small a quantity that we may consider the force of gravity for small elevations above the surface of the earth as practically constant. But it is variable for different points on the surface, being least at the equator, and gradually increasing as the latitude increases, according to a law which is approximately expressed by the formula

$$g = 32.1726 - 0.08238 \cos 2L,$$

in which $L =$ the latitude of the place,
$\quad\quad g =$ the acceleration due to gravity at the latitude L,
$\quad\quad\quad$ or simply the force of gravity, and
32.1726 ft. $=$ the force of gravity at latitude 45 degrees.

From this we find that

at the equator $g = g_0 = 32.09022$ feet, and
at the poles $g = g_{90} = 32.25498$ feet.

The varying force of gravity is determined by means of a pendulum, as will be shown hereafter. It is impossible to determine the exact law of relation between the force of gravity at different points on the surface of the earth, for it is not homogeneous nor an exact ellipsoid of revolution. The delicate observations made with the pendulum show that any assumed formula is subject to a small error. (See *Mécanique Céleste*, and *Puissant's Géodésie*.)

Substituting g for f in equations (11), (12), and (13), we have the following equations, which are applicable to bodies falling freely in *vacuo*:—

$$\left. \begin{array}{l} v = gt = \sqrt{2gs} = \dfrac{2s}{t}; \\[6pt] s = \tfrac{1}{2}gt^2 = \dfrac{v^2}{2g} = \tfrac{1}{2}vt; \\[6pt] t = \dfrac{v}{g} = \sqrt{\dfrac{2s}{g}} = \dfrac{2s}{v}. \end{array} \right\} \quad (16)$$

EXAMPLES.

1. A body falls through a height of 200 feet; required the time of descent and the acquired velocity. Let $g = 32\tfrac{1}{6}$ feet.

Ans. $t = 3.53$ seconds.
$v = 113.31$ feet.

2. A body is projected upward with a velocity of 1000 feet per second; required the height of ascent when it is brought to rest by the force of gravity.

Ans. $s = 15,544$ feet, nearly.

3. A body is dropped into a well and four seconds afterwards it is heard to strike the bottom. Required the depth, the velocity of sound being 1130 feet per second.

Ans. 231 feet.

4. A body is projected upward with a velocity of 100 feet per second, and at the same instant another body is let fall from a height 400 feet above the other body; at what point will they meet?

5. With what velocity must a body be projected downward that it may in t seconds overtake another body which has already fallen through a feet?

$$\text{Ans. } v = \frac{a}{t} + \sqrt{2ag}.$$

6. Required the space passed over by a falling body during the nth second.

20. Mass is quantity of matter. If we conceive that a quantity of matter, say a cubic foot of water, earth, stone, or other substance, is transported from place to place, without expansion or contraction, the quantity will remain the same, while its weight may constantly vary. If placed at the centre of the earth it will weigh nothing; if on the moon it will weigh less than on the earth, if on the sun it will weigh more; and if at any place in the universe its weight will be directly as the attractive force of gravity, and since the acceleration is also directly as the force of gravity, we have

$$\frac{W}{g} = \text{constant},$$

for the same mass at all places. This ratio for any contemporaneous values of W and g may be taken as the measure of the *mass*, as will be shown in the two following articles. The weight in these cases must be determined by a spring balance or its equivalent.

21. Dynamic measure of a force. Conceive that a body is perfectly free to move in the direction of the applied force, and that a constant *uniform* force, which acts either as a *pull* or *push*, is applied to the body. It will at the end of one second produce a certain velocity, which call $v_{(1)}$. If now forces of different intensities be applied to the same body they will produce velocities in the same time which are proportional to the forces; or

$$f \propto v_{(1)},$$

in which f is the applied force.

Again, if the same forces are applied to bodies having different masses, producing the same velocities in one second, then will the forces vary directly as the masses, or,

$$f \propto M.$$

Hence, generally, if *uniform, constant* forces are applied to different masses producing velocities $v_{(1)}$ in one second, then

$$f \propto Mv_{(1)};$$

or, in the form of an equation,

$$f = cMv_{(1)}; \qquad (17)$$

where c is a constant to be determined.

If the forces are *constantly varying*, the velocities generated at the end of one second will not measure the intensities at any instant, but according to the above reasoning, the *rate of variation of the velocity* will be one of the elements of the measure of the force. Hence if

$F =$ a *variable* force;
$M =$ the mass moved;
$\dfrac{dv}{dt} = f =$ the rate of variation of the velocity; or

velocity-increment;

and, $\dfrac{dv}{dt}$ be substituted for $v_{(1)}$ in equation (17), reducing by equation (6), we have

$$F = cMf = cM\frac{dv}{dt} = cM\frac{d^2s}{dt^2}. \qquad (18)$$

From this we have

$$cM = \frac{F}{f};$$

hence the value of cM is expressed in terms of the constant ratio of the force F to that of the acceleration f.

To determine this ratio experimentally I suspended a weight, W, by a very long fine wire. The wire should be long, so that the body will move practically in a straight line for any arc through which it will be made to move, and it should be very small, so that it will contain but little mass. By means of suitable mechanism I caused a constant force, F, to be applied horizontally to the body, thus causing it to move sidewise, and determined

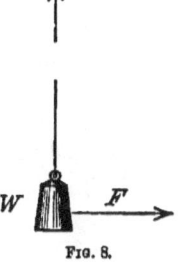

FIG. 8.

the space over which it passed during the first second. This equalled one-half the acceleration (see the first of equations (12) when $t = 1$). I found when $F = \frac{1}{20} W$, that $f = 1.6$ feet, nearly; and for $F = \frac{1}{10} W$, $f = 3.2$ feet, nearly; and similarly for other forces; hence

$$cM = \tfrac{1}{32} W, \text{ nearly.}$$

But the ratio of F to f is determined most accurately and conveniently by means of falling bodies; for $f = g =$ the acceleration due to the force of gravity, and W the weight of the body (which is a measure of the statical effect of the force of gravity upon the body), hence

$$cM = \frac{W}{g}; \qquad (19)$$

in which the values of W and g must be determined at the same place; but that place may be anywhere in the universe. The value of c is assumed, or the relation between c and M fixed arbitrarily.

If $c = 1$, we have

$$M = \frac{W}{g}; \qquad (20)$$

and this is the expression for the *mass*, which is nearly if not quite universally adopted. This in (18) gives

$$F = M \frac{d^2 s}{dt^2} = \frac{W}{g} \frac{d^2 s}{dt^2}; \qquad (21)$$

and hence THE DYNAMIC MEASURE OF THE PRESSURE WHICH MOVES A BODY *is the product of the mass into the acceleration*. This is sometimes called an accelerating force.

If there are retarding forces, such as friction, resistance of the air or water, or forces pulling in the opposite direction; then the first member F, is the measure of the unbalanced forces in pounds, and the second member is its dynamic equivalent.

22. UNIT OF MASS. If it is assumed that $c = 1$, as in the preceding article, the *unit of mass* is virtually fixed. In (20) if $W = 1$ and $g = 1$, then $M = 1$; that is, *a unit of mass* is the quantity of matter which will weigh one pound at that place in the universe where the acceleration due to gravity is one. If a quantity of matter weighs $32\tfrac{1}{6}$ lbs. at a place where $g = 32\tfrac{1}{6}$ feet, we have

$$M = \frac{32\tfrac{1}{6}}{32\tfrac{1}{6}} = 1;$$

hence on the surface of the earth a body which weighs $32\tfrac{1}{6}$ pounds (nearly) is *a unit of mass*.

It would be an *exact* unit if the acceleration were exactly $32\tfrac{1}{6}$ feet.

In order to illustrate this subject further, suppose that we make the *unit of mass* that of a *standard pound*. Then equation (19) becomes

$$c \cdot 1 = \frac{1}{g_0},$$

in which g_0 is the value of g at the latitude of 45 degrees. This value resubstituted in the same equation gives

$$M = \frac{g_0}{g} W,$$

and these values in equation (18) give

$$F = \frac{1}{g_0} Mf = \frac{W}{g} \frac{d^2 s}{dt^2};$$

the final value of which is the same as (21).

Again, if the *unit of mass* were the weight of one cubic foot of distilled water at the place where $g_0 = 32.1801$ feet, at which place we would have $W = 62.3791$, and (19) would give

$$c \cdot 1 = \frac{62.3791}{32.1801},$$

and this in the same equation gives

$$M = \frac{32.1801}{62.3791} \cdot \frac{W}{g},$$

and these values in (18) give

$$F = \frac{W}{g} \frac{d^2 s}{dt^2}, \text{ as before.}$$

23. Density *is the mass of a unit of volume.*

If M = the mass of a body;
V = the volume; and
D = the density;

then if the density is uniform, we have

$$D = \frac{M}{V}.$$

If the density is variable, let

δ = the density of any element, then

$$\delta = \frac{dM}{dV};$$

$$\therefore M = \int \delta \, dV \qquad (22)$$

from which the mass may be determined when δ is a known function of V.

EXAMPLES.

1. In a prismatic bar, if the density increases uniformly from one end to the other, being zero at one end and 5 at the other, required the total mass.

Let l = the length of the bar;
A = the area of the transverse section; and
x = the distance from the zero end;

then will

$\dfrac{5}{l}$ = the density at a unit's distance from the zero end;

$\dfrac{5}{l}x$ = the density at a distance x; and

$dV = A\,dx;$

$$\therefore M = A \int_0^l \frac{5x\,dx}{l} = \frac{5}{2} Al.$$

2. In a circular disc of uniform thickness, if the density at a unit's distance from the centre is 2, and increases directly as the distance from the centre, required the mass when the radius is 10.

APPLICATIONS OF EQUATION (21).

3. In the preceding problem suppose that the density increases as the square of the radius, required the mass.

4. In the preceding problem if the density is two pounds per cubic foot, required the weight of the disc.

5. If in a cone, the density diminishes as the cube of the distance from the apex, and is *one* at a distance one from the apex, required the mass of the cone.

Having established a unit of density, we might properly say that mass is a certain number of *densities*.

24. APPLICATIONS OF EQUATION (21).

[OBS.—If, for any cause, it is considered desirable to omit any of the matter of the following article, the author urges the student to at least establish the equations for the acceleration for each of the 31 examples here given. This part belongs purely to mechanics. The reduction of the equations belongs to mathematics. It would be a good exercise to establish the fundamental equations for all these examples, before making any reductions. Such a course serves to impress the student with the distinction between mechanical and mathematical *principles*.]

1st. *If a body whose weight is* 50 *pounds is moved horizontally by a constant force of* 10 *pounds, required the velocity acquired at the end of* 10 *seconds and the space passed over during that time, there being no friction nor other external resistance, and the body starting from rest.*

Fig. 9.

Here

$$M = \frac{W}{g} = \frac{50}{32\tfrac{1}{6}} \text{ lbs., and}$$

$$F = 10 \text{ lbs.};$$

hence (21) gives

$$\frac{d^2s}{dt^2} = \frac{F}{M} = \frac{193}{30}.$$

Multiply by dt and integrate, and

$$\frac{ds}{dt} = v = \frac{193}{30} t + (C_1 = 0).$$

The second integral is

$$s = \frac{193}{60} t^2 + (C_2 = 0);$$

and hence for $t = 10$ seconds, we have

$$v = 64.33 + \text{ feet.}$$
$$s = 321.66 + \text{ feet.}$$

2d. Suppose the data to be the same as in the preceding example, and also that the friction between the body and the plane is 5 pounds. Required the space passed over in 10 seconds.

Here $F = (10 - 5)$ pounds.

$$\therefore \frac{d^2s}{dt^2} = \frac{F}{M} = \frac{193}{60}.$$

3d. Suppose that a body whose weight is 50 pounds is moved horizontally by a weight of 10 lbs., which is attached to an inextensible, but perfectly flexible string which passes over a wheel and is attached at the other end to the body. Required the distance passed over in 10 seconds, if the string is without weight, and no resistance is offered by the wheel, plane, or string.

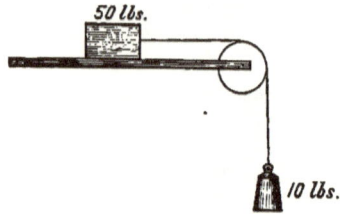

FIG. 10.

In this case gravity exerts a force of 10 pounds to move the mass, or $F = 10$ lbs., and the mass moved is that of both bodies, or $M = (50 + 10) \div 32\frac{1}{6}$.

$$\frac{d^2s}{dt^2} = \frac{F}{M} = \frac{193}{36}.$$

The integration is performed as before.

Ans. $s = 268.05$ feet.

4th. Find the tension of the string in the preceding example.

The tension will equal that *force* which, if applied directly to the body, as in Ex. 1, will produce the same acceleration as in the preceding example.

Let $P = 10$ pounds;
$W = 50$ pounds;
$T =$ tension;
$\dfrac{P + W}{g} =$ the mass in the former example; and
$\dfrac{W}{g} =$ the mass moved by the tension.

Hence, from Equation (21),

$$\frac{P + W}{g} f = P ; \text{ and}$$

$$\frac{W}{g} f = T.$$

Eliminate f, and we find

$$T = \frac{WP}{W + P};$$

$$\therefore T = 8.33 \text{ lbs.}$$

What must be the value of P so that the tension will be a maximum or a minimum, $P + W$ being constant?

5th. In example 3, what must be the weight of P so that the tension shall be $(\frac{1}{n})^{\text{th}}$ part of P?

Ans. $P = (n-1) W$.

6th. If a body whose weight is W falls freely in a vacuum by the force of gravity, determine the formulas for the motion.

Here $Mg = W$ and the moving force $F = W$;

$$\therefore \frac{W}{g} \frac{d^2s}{dt^2} = W;$$

$$\text{or, } \frac{d^2s}{dt^2} = g.$$

The integrals of this equation will give Equations (16), when the initial space and velocity are zero. Let the student deduce them.

7th. Suppose that the moving pressure (pull or push) equals the weight of the body, required the velocity and space.

Here $Mg = W$ and $F = W$, hence the circumstances of motion will be the same as in the preceding example.

The forces of nature produce motion without *apparent* pressure, but this example shows that their effect is the same as that produced by a push or pull whose intensity equals the weight of the body, and hence both are measured by *pounds*, or their equivalent.

8th. If the force F is constant, show that $Ft = Mv$; also $Fs = \tfrac{1}{2}Mv^2$, and $\tfrac{1}{2}Ft^2 = Ms$. If F is variable we have $Mv = \int F dt.$

9th. Suppose that a piston, devoid of friction, is driven by a constant steam-pressure through a portion of the length of a cylinder, at what point in the stroke must the pressure be instantly reversed so that the full stroke shall equal the length of the cylinder, the cylinder being horizontal?

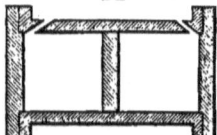

Fig. 11.

At the middle of the stroke. Whatever velocity is generated through one-half the stroke will be destroyed by the counter pressure during the other half.

10th. If the pressure upon the piston is 500 pounds, weight of the piston 50 pounds, and the friction of the piston in the cylinder 100 pounds, required the point in the stroke at which the pressure must be reversed that the stroke may be 12 inches.

The uniform effective pressure for driving the piston is $500 - 100 = 400$ lbs., and the uniform effective force for stopping the motion is $500 + 100 = 600$ pounds. The velocity generated equals the velocity destroyed, and the velocity destroyed equals that which would be generated in the same space by a force equal to the resisting force; hence if

$F =$ the effective moving force;
$s =$ the space through which it acts;
$v =$ the resultant velocity;
$F' =$ the resisting force; and
$s' =$ the space through which it acts;

then, from the expression in Example 8, we have
$$Fs = \tfrac{1}{2}Mv^2,$$
and
$$F's' = \tfrac{1}{2}Mv^2,$$
$$\therefore Fs = F's',$$
or,
$$F : F' :: s' : s.$$

In the example, $F = 400$ lbs., and $F' = 600$ lbs. Let $x =$ the distance from the starting point to the point where the pressure must be reversed. Then
$$600 : 400 :: x : 12 - x, \therefore x = 7\tfrac{1}{5} \text{ inches.}$$

11*th.* *If in the preceding example the piston moves vertically up and down, required the point at which the pressure must be instantly reversed so that the full stroke shall be 12 inches.*

The effective driving pressure upward will be $500 - 100 - 50 = 350$ pounds, and the retarding force will be $500 + 100 + 50 = 650$ pounds, and during the down-stroke the driving force is $500 + 50 - 100 = 450$ pounds, and the retarding force is $500 - 50 + 100 = 550$ pounds.

12*th.* *A string passes over a wheel and has a weight P attached at one end, and W at the other. If there are no resistances from the string or wheel, and the string is devoid of weight, required the resulting motion.*

Suppose $W > P$; then

FIG. 12.

$$F = W - P, \text{ and}$$
$$M = \frac{W + P}{g};$$
$$\therefore \frac{d^2s}{dt^2} = \frac{F}{M} = \frac{W - P}{W + P} g.$$

By integrating, we find
$$v = \frac{W - P}{W + P} gt,$$
and,
$$s = \tfrac{1}{2} \frac{W - P}{W + P} gt^2.$$

13*th.* *Required the tension of the string in the preceding example.*

The tension equals the weight P, plus the force which will produce the acceleration

$$\frac{W-P}{W+P} g$$

when applied to raise P vertically. The mass multiplied by the acceleration is this moving force, or

$$\frac{P}{g} \cdot \frac{W-P}{W+P} g;$$

hence the tension is

$$P + \frac{W-P}{W+P} P = \frac{2WP}{W+P}.$$

Similarly, it equals W minus the accelerating force, or

$$W - \frac{W-P}{W+P} W = \frac{2WP}{W+P}.$$

A complete solution of this class of problems involves the mass of the wheel and frictions, and will be considered hereafter.

14*th.* *A string passes over a wheel and has a weight P attached to one end and on the other side of the wheel is a weight W, which slides along the string. Required the friction between the weight W and the string, so that the weight P will remain at rest. Also required the acceleration of the weight W.*

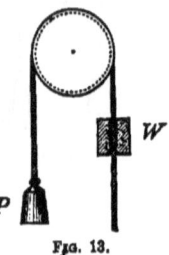
Fig. 13.

The friction $= P$;

$$Mg = W;$$
$$\text{and, } F = W - P;$$
$$\therefore \frac{d^2s}{dt^2} = \frac{F}{M} = \frac{W-P}{W} g;$$

hence,

$$v = \frac{W-P}{W} gt,$$

and,

$$s = \tfrac{1}{2} \frac{W-P}{W} gt^2.$$

15th. *In the preceding example, if W were an animal whose weight is less than P, required the acceleration with which it must ascend, so that P will remain at rest.*

16th. *If the weight W descend along a rough rope with a given acceleration, required the acceleration with which the body P must ascend or descend on the opposite rope, so that the rope may remain at rest, no allowance being made for friction on the wheel.*

(The ascent must be due to climbing up on the cord, or be produced by an equivalent result.)

17th. *A particle moves in a straight line under the action of a uniform acceleration, and describes spaces s and s' in t^{th} and t'^{th} seconds respectively, determine the accelerating force and the velocity of projection.*

Let $v_0 =$ the velocity of projection, and
$f =$ the acceleration;

then
$$f = \frac{s' - s}{t' - t},$$

and
$$v_0 = \frac{s'(2t - 1) - s(2t' - 1)}{2(t - t')}.$$

If $\frac{s'}{s} = \frac{2t' - 1}{2t - 1}$, then $v_0 = 0$.

18th. *If a perfectly flexible and perfectly smooth rope is placed upon a pin, find in what time it will run itself off.*

If it is perfectly balanced on the pin it will not move, unless it receive an initial velocity. If it be unbalanced, the weight of the unbalanced part will set it in motion. Suppose that it is balanced and let

$v_0 =$ the initial velocity,
$2l =$ the length of the rope,
$w =$ the weight of a unit of length, and
$t =$ the time.

Take the origin of coördinates at the end of the rope at the instant that motion begins. When one end has descended *s* feet, the other has ascended the same amount, and hence the unbalanced weight will be $2ws$. The mass moved will be $2wl \div g$; hence we have

$$\frac{d^2s}{dt^2} = \frac{F}{M} = \frac{2\,ws}{2\,wl}g = \frac{g}{l}s.$$

Multiply by ds and integrate, and we have

$$\frac{ds^2}{dt^2} = v^2 = \frac{g}{l}s^2 + (C = v_0^2);$$

$$\therefore \sqrt{\frac{g}{l}}dt = \frac{ds}{\sqrt{\frac{l}{g}v_0^2 + s^2}};$$

Integrating again, gives

$$t = \sqrt{\frac{l}{g}} \log \left\{ \frac{s + \sqrt{\frac{l}{g}v_0^2 + s^2}}{\sqrt{\frac{l}{g}}v_0} \right\}$$

$$= \sqrt{\frac{l}{g}} \log \left\{ \frac{l + \sqrt{\frac{l}{g}v_0^2 + l^2}}{\sqrt{\frac{l}{g}}v_0} \right\}, \text{ if } s = l.$$

19th. *If a particle moves towards a centre of force which* ATTRACTS *directly as the distance from the force, determine the motion.*

Let μ = the absolute force; that is, the acceleration at a unit's distance from the centre due to the force; and

s = the distance; .

then

$$\frac{d^2s}{dt^2} = -\mu s.$$

The second member is negative, because s is an inverse function of t, that is, as t increases s diminishes. *Generally attractive forces are negative and repulsive ones positive*, in reference to the centre of the force. This is the same equation as in Example 5, Article 17; hence, if a is the initial value of s,

$$v = \sqrt{\mu (a^2 - s^2)};$$

and

$$t = \mu^{-\frac{1}{2}} (\sin^{-1} \frac{s}{a} - \tfrac{1}{2}\pi);$$

and the velocity at the centre of the force is found by making $s = 0$, for which we have,

$$v = a \sqrt{\mu};$$

and $\quad t = -\dfrac{1}{2}\mu^{-\frac{1}{2}}\pi,\ \dfrac{1}{2}\mu^{-\frac{1}{2}}\pi,\ \dfrac{3}{2}\mu^{-\frac{1}{2}}\pi,$ etc.,

hence, the time is independent of the initial distance.

It may be proved that within a homogeneous sphere the attractive force varies directly as the distance from the centre. Hence, if the earth were such a sphere, and a body were permitted to pass freely through it, it would move with an accelerated velocity from the surface to the centre, at which point the velocity would be a maximum, and it would move on with a retarded velocity and be brought to rest at the surface on the opposite side. It would then return to its original position, and thus move to and fro, like the oscillations of a pendulum.

The acceleration due to gravity at the surface of the earth being g, and r being the radius, the absolute force is

$$\mu = \frac{g}{r};$$

$$\therefore v = r\sqrt{\mu} = \sqrt{g.r};$$

and the time of passing from surface to surface on the equator would be

$$t = \pi\sqrt{\frac{r}{g}} = 3.1416\sqrt{\frac{20{,}923{,}161}{32.09025}} = 42\text{m. }1.6\text{ sec.}$$

The *exact* dimensions of the earth are unknown. The semi-polar axis of the earth is, as determined by

Bessel............................ 20,853,662 ft.
Airy............................. 20,853,810 ft.
Clarke .... 20,853,429 ft.

The equatorial radius is not constant, on account of the elevations and depressions of the surface. There are some indications that the general form of the equator is an ellipse. Among the more recent determinations are those by Mr. Clarke, of England (1873), and his result given below is considered by him as the most probable *mean*. The equatorial radius, is according to

Bessel .... 20,923,596 ft.
Airy........................... 20,923,713 ft.
Clarke......................... 20,923,161 ft.

The determination of the force of gravity at any place is subject to small errors, and when it is *computed* for different places the result may differ from the actual value by a perceptible amount.

The force of gravity at any particular place is *assumed* to be constant, but all we can assert is that if it is variable the most delicate observations have failed to detect it. But it is well known that the surface of the earth is constantly undergoing changes, being elevated in some places and depressed in others, and hence, assuming the law of gravitation to be exact and universal, we cannot escape the conclusion that the force of gravity at every place on its surface changes, and although the change is exceedingly slight, and the total change may extend over long periods of time, it may yet be possible, with apparatus vastly more delicate than that now used, to measure this change. It seems no more improbable than the solution of many problems already attained—such for instance, as determining the relative velocities of the earth and stars by means of the spectroscope.

20th. Suppose that a coiled spring whose natural length is A B, is compressed to B C. If one end rests against an immovable body B, and the other against a body at C, which is perfectly free to move horizontally, what will be the time of movement from C to A, and what will be the velocity at A ?

FIG. 14.

It is found by experiment that the resistance of a spring to compression varies directly as the amount of compression, hence the action of the spring in pushing the body, will, in reference to the point A, be the same as an attractive force which varies directly as the distance, and hence it is similar to the preceding example. But if the spring is not attached to the particle the motion will not be periodic, but when the particle has reached the point A it will leave the spring and proceed with a uniform velocity. If the spring were destitute of mass, it would extend to A, and become instantly at rest, but because of the mass in it, the end will pass A and afterwards recoil and have a periodic motion. If the body be *attached* to the spring, it will have a periodic motion, and the solution will be similar to the one in the Author's *Resistance of Materials*, Article 19.

Take the origin at A, s being counted to the left; suppose that 5 pounds will compress the spring one inch, and let the total compression be $a = 4$ inches. Let $W =$ the weight of the body $= 10$ pounds.

[24.]   ACCELERATING FORCES.   31

The force at the distance of one foot from the origin being 60 pounds, the force at s feet will be $60s$ pounds.

Hence, $$\frac{W}{g}\frac{d^2s}{dt^2} = -F = -60s;$$

or, $$\frac{d^2s}{dt^2} = -193s;$$

from which we find that

$$t = \frac{\frac{1}{2}\pi}{\sqrt{193}},$$

and $v = 4.6 +$ feet.

21st. *Suppose that in the preceding problem a body whose weight is M' is at B, and another M'' at C, both being perfectly free to move horizontally, required the time of movement that the distance between them shall be equal to AB; and the resultant velocities of each.*

It is convenient in this case to take the origin at the centre of one of the bodies—say that of B—and remain at the centre during motion. The origin will be movable.

Let $\mu =$ the absolute force of the spring; that is, the force which will compress the spring a unit of length —say one inch—;

$a =$ the total compression; and

$b =$ the length after compression.

Then

$\mu a =$ the total reaction of the spring when motion begins;

$\mu(a-(s-b)) =$ the reaction (or moving force) when the spring has expanded an amount equal to $s - b$; and

$M' + M'' =$ the total mass moved;

hence

$$\frac{d^2s}{dt^2} = \frac{F}{M} = \frac{\mu(a+b-s)}{M'+M''};$$

$$\therefore \frac{ds^2}{dt^2} = \frac{\mu(2as + 2bs - s^2)}{M' + M''} + C_1.$$

But $v = 0$ for $s = b$; $\therefore C_1 = -\mu \dfrac{(2a+b)b}{M'+M''}$;

$\therefore \sqrt{\dfrac{\mu}{M'+M''}}\, dt = \dfrac{ds}{\sqrt{2(s-b)a - (s-b)^2}}$;

and integrating gives

$$\sqrt{\dfrac{\mu}{M'+M''}}\, t = \text{vers}^{-1} \dfrac{s-b}{a} + (C_2 = 0);$$

and, making $s = a + b$, we have

$$t = \tfrac{1}{2}\pi \sqrt{\dfrac{M'+M''}{\mu}};$$

which, as in the preceding example, is independent of the amount of compression of the spring.

To find the relation between the absolute velocities,

Let s' = the space passed over by M', and
s'' = the space passed over by M'';

then since the moving force is the same for both, we have

$$M' \dfrac{d^2 s'}{dt^2} = M'' \dfrac{d^2 s''}{dt^2}.$$

Integrating, gives

$$M'v' = M''v''.$$

22d. *Suppose that the force varies directly as the distance from the centre of force and is* REPULSIVE.
Then

$$\dfrac{d^2 s}{dt^2} = \mu s;$$

$$\therefore s = \dfrac{v_0}{2\sqrt{\mu}}\left(e^{t\sqrt{\mu}} - e^{-t\sqrt{\mu}}\right),$$

in which v_0 is the initial velocity.

23d. *Suppose that the force varies inversely as the square of the distance from the centre of the force and is* ATTRACTIVE.

[This is the law of universal gravitation, and is known as the law of the *inverse squares*. While it is rigidly true, so far as we know, for every particle of matter acting upon any other particle, it is not rigidly true for finite bodies acting upon other bodies at a finite distance, except for *homogeneous spheres*, or spheres composed of homogeneous shells. The earth being neither homogeneous nor a sphere, it will not be *exactly* true that it attracts external bodies with a force which varies inversely as the square of the distance from the centre, but the deviations from the law for bodies at great distances from the earth will not be perceptible. We assume that the law applies to all bodies above the surface of the earth, the centre of the force being at the centre of the earth.]

Let the problem be applied to the attraction of the earth, and

$r =$ the radius of the earth;
$g =$ the force of gravity at the surface;
$\mu =$ the absolute force; and
$s =$ the distance from the centre;

then

$$\mu = gr^2;$$

and

$$\frac{d^2s}{dt^2} = -\frac{\mu}{s^2}.$$

Multiply by ds and integrate; observing that for $s = a$, $v = 0$, and we have

$$\frac{ds^2}{dt^2} = 2\mu \left(\frac{1}{s} - \frac{1}{a}\right) \qquad (a)$$

$$= 2\mu \frac{as - s^2}{as^2};$$

$$\therefore \left(\frac{2\mu}{a}\right)^{\frac{1}{2}} dt = \frac{-s\,ds}{(as - s^2)^{\frac{1}{2}}};$$

using the negative sign, because t and s are inverse functions of each other.

The second member may be put in a convenient form for integration by adding and subtracting $\frac{1}{2}a$ to the numerator and arranging the terms. This gives

$$\frac{\frac{1}{2}a - s - \frac{1}{2}a}{(as - s^2)^{\frac{1}{2}}} ds$$

$$= \frac{a - 2s}{2(as - s^2)^{\frac{1}{2}}} ds - \frac{a\,ds}{2(as - s^2)^{\frac{1}{2}}};$$

the integral of which is

$$(as - s^2)^{\frac{1}{2}} - \tfrac{1}{2}a \text{ versin}^{-1} \frac{2s}{a} + C.$$

But when $s = a, t = 0 \therefore C = \tfrac{1}{2}a\pi$;

$$\therefore t = \left(\frac{a}{2\mu}\right)^{\frac{1}{2}} \left\{ \left(as - s^2\right)^{\frac{1}{2}} + a \cos^{-1}\left(\frac{s}{a}\right)^{\frac{1}{2}} \right\} \qquad (b)$$

[From the circle we have $\pi - \text{versin}^{-1} \dfrac{2s}{a} = \pi - \cos^{-1}\left(1 - \dfrac{2s}{a}\right) =$

$\cos^{-1}\left(-\left(1 - \dfrac{2s}{a}\right)\right) = \cos^{-1}\left(\dfrac{2s}{a} - 1\right)$.

From trigonometry we have $2 \cos^2 y - 1 = \cos 2y$.

Let $2y = \cos^{-1}\left(\dfrac{2s}{a} - 1\right)$, then

$\cos 2y = \dfrac{2s}{a} - 1$; $\therefore \cos^2 y = \dfrac{s}{a}$, and $y = \cos^{-1}\sqrt{\dfrac{s}{a}}$; and

$2y = 2\cos^{-1}\sqrt{\dfrac{s}{a}}$; or $\pi - \text{versin}^{-1}\dfrac{2s}{a}$.].

From (a) it appears that for $s = 0, v = \infty$; hence the velocity at the centre will be infinite when the body falls from a finite distance.

If $s = a = \infty$, $v = 0$. If a body falls freely from an infinite distance to the earth, we have in equation (a)

$a = \infty$; and
$s = r =$ the radius of the earth ;

$$\therefore v = \left(\frac{2\mu}{r}\right)^{\frac{1}{2}},$$

for the velocity at the surface. But $\dfrac{\mu}{r^2} = g$;

$$\therefore v = (2gr)^{\frac{1}{2}}.$$

If $g = 32\tfrac{1}{6}$ feet and $r = 3962$ miles, we have

$$v = \left(\frac{64\tfrac{1}{3} \times 3962}{5280}\right)^{\frac{1}{2}} = 6.95 \text{ miles}.$$

Hence the maximum velocity with which a body can reach the earth is less than seven miles per second.

24th. *Suppose that the force is* ATTRACTIVE *and varies inversely as the nth power of the distance.*

Then
$$\frac{d^2s}{dt^2} = -\frac{\mu}{s^n};$$

$$\therefore \frac{ds^2}{dt^2} = \frac{2\mu}{n-1}\left(\frac{1}{s^{n-1}} - \frac{1}{a^{n-1}}\right);$$

and integrating, gives

$$t = \left(\frac{n-1}{2\mu}\right)^{\frac{1}{2}} a^{\frac{1}{2}(n-1)} \int_a^x (a^{n-1} - s^{n-1})^{-\frac{1}{2}} s^{\frac{1}{2}(n-1)} ds.$$

According to the *tests of integrability* this may be integrated when

$$n = \ldots \ldots \frac{5}{7}, \frac{3}{5}, \frac{1}{3}, -1, \frac{3}{11}, \text{ or } \frac{5}{3} \ldots \text{ etc.,}$$

$$\text{or } n = \ldots \ldots \frac{3}{4}, \frac{2}{3}, \frac{1}{2}, 0, 2, \text{ or } \frac{3}{2} \ldots \ldots \text{ etc.}$$

25th. *Let the force vary inversely as the square root of the distance and be* ATTRACTIVE. (This is one of the special cases of the preceding example.)

We have
$$\frac{d^2s}{dt^2} = -\frac{\mu}{s^{\frac{1}{2}}};$$

$$\therefore \frac{ds^2}{dt^2} = 4\mu(a^{\frac{1}{2}} - s^{\frac{1}{2}});$$

or, $2\mu^{\frac{1}{2}} dt = \dfrac{-ds}{(a^{\frac{1}{2}} - s^{\frac{1}{2}})^{\frac{1}{2}}}.$

The negative sign is taken because t and s are inverse functions of each other.

Add and subtract $\dfrac{2\sqrt{a}}{3\sqrt{s}\sqrt{a^{\frac{1}{2}} - s^{\frac{1}{2}}}}$ and we have

$$2\sqrt{\mu}\, dt = \left[\frac{\sqrt{s}}{\sqrt{s}\sqrt{a^{\frac{1}{2}} - s^{\frac{1}{2}}}} + \frac{2\sqrt{a}}{3\sqrt{s}\sqrt{a^{\frac{1}{2}} - s^{\frac{1}{2}}}} - \frac{2\sqrt{a}}{3\sqrt{s}\sqrt{a^{\frac{1}{2}} - s^{\frac{1}{2}}}}\right] ds$$

$$= \left[\frac{3\sqrt{s}+2\sqrt{a}}{3\sqrt{s}\sqrt{a^{\frac{1}{2}}-s^{\frac{1}{2}}}} - \frac{2\sqrt{a}}{3\sqrt{s}\sqrt{a^{\frac{1}{2}}-s^{\frac{1}{2}}}} \right] ds.$$

$$\therefore t = \frac{2}{3\sqrt{\mu}} \left[s^{\frac{1}{2}}(a^{\frac{1}{2}}-s^{\frac{1}{2}})^{\frac{1}{2}} + 2\sqrt{a}\,(a^{\frac{1}{2}}-s^{\frac{1}{2}})^{\frac{1}{2}} \right]$$

$$= \frac{2}{3\sqrt{\mu}}\left(s^{\frac{1}{2}} + 2a^{\frac{1}{2}}\right)\left(a^{\frac{1}{2}}-s^{\frac{1}{2}}\right)^{\frac{1}{2}}.$$

26th. *Suppose that the force is* ATTRACTIVE *and varies inversely as the distance.*

Hence

$$\frac{d^2s}{dt^2} = -\frac{\mu}{s};$$

$$\therefore \frac{ds^2}{dt^2} = 2\mu \log\frac{a}{s};$$

in which $s = a$ for $v = 0$. Hence the time from $s = a$ to $s = 0$, is

$$t = \frac{1}{\sqrt{2\mu}} \int_a^0 \frac{ds}{\left(\log\frac{a}{s}\right)^{\frac{1}{2}}} = a\left(\frac{\pi}{2\mu}\right)^{\frac{1}{2}}.$$

Let $\left(\log\frac{a}{s}\right)^{\frac{1}{2}} = y$; then for $s = a$, $y = 0$ and for $s = 0$, $y = \infty$. Squaring and passing to exponentials, we have

$$\log\frac{a}{s} = y^2 \quad \therefore \frac{a}{s} = e^{y^2}, \text{ or } s = a\,e^{-y^2};$$

$$\therefore ds = -a e^{-y^2}.\,2y\,dy;$$

$$\therefore t = a\left(\frac{2}{\mu}\right)^{\frac{1}{2}} \int_0^\infty e^{-y^2}\,dy = a\left(\frac{\pi}{2\mu}\right)^{\frac{1}{2}}.$$

This is called a *gamma-function*, and a method of integrating it is as follows:—

Since functions of the same *form* integrated between the same limits are independent of the variables and have the same value, therefore

[24.] ATTRACTIVE FORCES. 37

$$\int_0^\infty e^{-y^2} dy = \int_0^\infty e^{-t^2} dt;$$

and $$\int_0^\infty e^{-y^2} dy \int_0^\infty e^{-t^2} dt = \left[\int_0^\infty e^{-y^2} dy\right]^2.$$

Also the left hand member will be of the same value if the sign of integration be placed over the whole of it, since the actual integration will be performed in the same order; hence

$$\left[\int_0^\infty e^{-y^2} dy\right]^2 = \int_0^\infty \int_0^\infty e^{-y^2-t^2} dy\, dt$$

$$= \int_0^\infty \int_0^\infty t e^{-t^2(1+u^2)} dt\, du;$$

in which $y = tu$; $\therefore dy = t\, du$. Integrating in reference to t, we have

$$\frac{-e^{-t^2(1+u^2)}}{2(1+u^2)} du,$$

which for $t = \infty$ becomes zero, and for $t = 0$ becomes $\dfrac{du}{2(1+u^2)}$, and the integral of this is $\tfrac{1}{2}\tan^{-1} u$, which is zero for $u = 0$, and $\tfrac{1}{4}\pi$ for $u = \infty$;

$$\therefore \int_0^\infty e^{-y^2} dy = \frac{1}{2}\sqrt{\pi}.$$

(See also *Méc. Céleste*, p. 151 [1534 O].)

Or we may proceed as follows:—

Let $e^{-t^2} = x \therefore dt = -(-\log x)^{\frac{1}{2}} \dfrac{dx}{2x};$

$$\therefore \int_0^\infty e^{-t^2} dt = \int_1^0 -\tfrac{1}{2}(-\log x)^{-\frac{1}{2}} dx.$$

Let $x = a^{y^2}$ and consider a less than unity; then $\log a$ will be negative, and $-\log x = y^2 \log a$;

$$\therefore dx = -a^{y^2} 2y\, dy \log a;$$

which substituted above gives

$$\int_0^\infty e^{-t^2} = \int_1^0 -\tfrac{1}{2}(-\log x)^{-\frac{1}{2}}dx = (-\log a)^{-\frac{1}{2}} \int_0^\infty a^{y^2} dy.$$

Dividing by $(-\log a)^{\frac{1}{2}}$ and multiplying both sides by $-da$, we have

$$\int_1^0 -\tfrac{1}{2}(-\log a)^{-\frac{1}{2}}da \int_1^0 -\tfrac{1}{2}(-\log x)^{-\frac{1}{2}}dx = \int_0^\infty \int_1^0 -\tfrac{1}{2}a^{y^2} dy\, da.$$

Integrating the second member first in regard to a, gives

$$-\tfrac{1}{2}\frac{a^{y^2+1}}{y^2+1}dy\,;$$

which between the limits of 0 and 1 gives $\tfrac{1}{2}\dfrac{dy}{1+y^2}$; the integral of which is $\tfrac{1}{2}\tan^{-1}y$ which between the limits of ∞ and 0 gives $\tfrac{1}{4}\pi$.

$$\therefore \int_1^0 -\tfrac{1}{2}(-\log a)^{-\frac{1}{2}}\,da \int_1^0 -\tfrac{1}{2}(-\log x)^{-\frac{1}{2}}dx$$

$$= \left[\int_1^0 -\tfrac{1}{2}(-\log x)^{-\frac{1}{2}}dx\right]^2 = \tfrac{1}{4}\pi\,;$$

$$\therefore \int_0^\infty e^{-t^2}\,dt = \tfrac{1}{2}\sqrt{\pi}.$$

(See *Méc. Céleste*, Vol. iv. p. 487, Nos. [8319] to [8331]. Chauvenet's *Spherical Astronomy*, Vol. i. p. 152. Todhunter's *Integral Calculus*. Price's *Infinitesimal Calculus*.)

Sometimes the best way to integrate an exponential quantity, is first to differentiate a similar one, and the integration often becomes apparent. Thus to integrate $t\,e^{-t^2}\,dt$, first differentiate e^{-t^2}. We have $d\,e^{-t^2} = e^{-t^2}$ $d(-t^2) = e^{-t^2}(-2\,tdt) = -2te^{-t^2}dt\,;$

$$\therefore \int d e^{-t^2} = -2\int t e^{-t^2} dt.$$

But the first member is the integral of the differential, and hence is the quantity itself, or e^{-t^2}, and hence the required integral is $-\tfrac{1}{2}\,e^{-t^2}$.

27th. *Suppose that two bodies have their centres at A and A' respectively, and attract a particle at p with forces which vary as the distances from A and A'.*

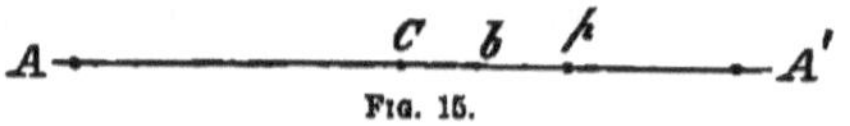

Fig. 15.

Let C be midway between A and A';
$Cp = c$;
$AC = CA' = a$;
$Cb = s =$ any variable distance;
and let $\mu = \mu' =$ the absolute forces of the bodies A and A' respectively.

Then

$$\frac{d^2s}{dt^2} = \mu(a-s) - \mu(a+s) = -2\mu s;$$

$$\therefore \frac{ds^2}{dt^2} = 2\mu(c^2 - s^2);$$

and integrating again gives

$$s = c \cos t \sqrt{2\mu}.$$

28th. *Suppose that a particle is projected with a velocity u into a medium which resists as the square of the velocity; determine the circumstances of motion.*

Take the origin at the point of projection, and the axis s to coincide with the path of the body.

Let $\mu =$ the absolute resistance—or the resistance of the medium when the velocity is unity;

then $\mu \left(\frac{ds}{dt}\right)^2 =$ the resistance for any velocity;

$$\therefore \frac{d^2s}{dt^2} = -\mu\left(\frac{ds}{dt}\right)^2;$$

or,

$$\frac{d\left(\frac{ds}{dt}\right)}{\frac{ds}{dt}} = -\mu ds.$$

And integrating between the initial limits, $s = 0$ for $\frac{ds}{dt} = u$, and the general limits, we have

$$\log \frac{ds}{dt} - \log u = -\mu s;$$

or,
$$\log \frac{\frac{ds}{dt}}{u} = -\mu s;$$

$$\therefore \frac{ds}{dt} = u e^{-\mu s};$$

or,
$$u\, dt = e^{\mu s} ds.$$

Integrating again, observing that $t = 0$ for $s = 0$, we have

$$\mu u t = e^{\mu s} - 1.$$

The velocity becomes zero only when $s = \infty$.

29th. *A heavy body falls in the air by the force of gravity, the resistance of the air varying as the square of the velocity; determine the motion.*

Take the origin at the starting point, and

Let $\kappa =$ the resistance of the body for a unit of velocity;
$s =$ the distance from the initial point, positive downwards;
$t =$ the time of falling through distance s;

then $\frac{ds}{dt} = 0$ for t and $s = 0$;

$\frac{ds}{dt} = v$ for $t = t$ and $s = s$; and

$\kappa \left(\frac{ds}{dt}\right)^2 =$ the resistance of the air at any point, and acts upwards;

and $g =$ the accelerating force downward;

hence, the resultant acceleration is the difference of the two, or

$$\frac{d^2 s}{dt^2} = g - \kappa \left(\frac{ds}{dt}\right)^2; \qquad (a)$$

or,
$$\frac{d\left(\frac{ds}{dt}\right)}{\kappa\, dt} = \frac{g}{\kappa} - \left(\frac{ds}{dt}\right)^2;$$

FALLING BODIES.

$$\therefore \kappa\, dt = \frac{d\left(\dfrac{ds}{dt}\right)}{\dfrac{g}{\kappa} - \left(\dfrac{ds}{dt}\right)^2}.$$

Separating this into two partial fractions, and integrating, gives

$$\kappa t = \tfrac{1}{2} \left(\frac{\kappa}{g}\right)^{\tfrac{1}{2}} \log \frac{g^{\tfrac{1}{2}} + \kappa^{\tfrac{1}{2}} \dfrac{ds}{dt}}{g^{\tfrac{1}{2}} - \kappa^{\tfrac{1}{2}} \dfrac{ds}{dt}}.$$

Passing to exponentials gives

$$v = \frac{ds}{dt} = \left(\frac{g}{\kappa}\right)^{\tfrac{1}{2}} \frac{e^{2t(\kappa g)^{\tfrac{1}{2}}} - 1}{e^{2t(\kappa g)^{\tfrac{1}{2}}} + 1}; \qquad (b)$$

which gives the velocity in terms of the time. To find it in terms of the space, multiply equation (a) by ds and put it under the form

$$\frac{d\left(\dfrac{ds}{dt}\right)^2}{\dfrac{g}{\kappa} - \left(\dfrac{ds}{dt}\right)^2} = 2\kappa\, ds.$$

Proceeding as before, observing the proper limits, we find

$$2\kappa s = -\log \frac{g - \kappa\left(\dfrac{ds}{dt}\right)^2}{g};$$

$$\therefore \frac{ds}{dt} = v = \sqrt{\frac{g}{\kappa}\left(1 - e^{-2\kappa s}\right)}. \qquad (c)$$

If $s = \infty$, $v = \sqrt{\dfrac{g}{\kappa}}$, and hence the velocity tends towards a constant.

From equation (b), multiplying the terms of the fraction by $e^{-t(\kappa g)^{\tfrac{1}{2}}}$, and observing that the numerator becomes the differential of the denominator, integrating, and passing to exponentials, we have,

$$2 e^{\kappa s} = e^{t(\kappa g)^{\tfrac{1}{2}}} + e^{-t(\kappa g)^{\tfrac{1}{2}}}; \qquad (d)$$

which gives the space in terms of the time.

A neat solution of equation (a) may be found by Lagrange's method of *Variation of Parameters*.

FALLING BODIES. [24.]

30th. Suppose that the body is projected upward in the air, having the same coefficient of resistance as in the preceding example.

Take the origin at the point of propulsion, u being the initial velocity; then

$$\frac{d^2s}{dt^2} \text{ or, } \frac{d}{dt}\frac{ds}{dt} = -g - \kappa\left(\frac{ds}{dt}\right)^2; \qquad (e)$$

hence,
$$\kappa dt = -\frac{d\left(\frac{ds}{dt}\right)}{\frac{g}{\kappa} + \left(\frac{ds}{dt}\right)^2};$$

$$\therefore \kappa t = -\left(\frac{\kappa}{g}\right)^{\frac{1}{2}}\left\{ \tan^{-1}\left(\frac{\kappa}{g}\right)^{\frac{1}{2}}\left(\frac{ds}{dt}\right) - \tan^{-1}\left(\frac{\kappa}{g}\right)^{\frac{1}{2}}u \right\}.$$

Solving this equation for $\frac{ds}{dt}$; we have,

$$v = \frac{ds}{dt} = \left(\frac{g}{\kappa}\right)^{\frac{1}{2}} \frac{u\sqrt{\kappa} - g \tan t\sqrt{\kappa g}}{\sqrt{g} + u\sqrt{\kappa}\tan t\sqrt{\kappa g}}. \qquad (f)$$

Substitute $\sin t\sqrt{\kappa g} + \cos t\sqrt{\kappa g}$ for $\tan t\sqrt{\kappa g}$ and the numerator becomes the differential of the denominator, and observing that $t = 0$ for $x = 0$, we have

$$s = \frac{1}{\kappa} \log \frac{u\sqrt{\kappa}\sin t\sqrt{\kappa g} + \sqrt{g}\cos t\sqrt{\kappa g}}{\sqrt{g}};$$

which gives the space in terms of the time.

Multiply equation (e) by ds and it may be put under the form

$$2\kappa ds = -\frac{d\left(\frac{ds}{dt}\right)^2}{\frac{g}{\kappa} + \left(\frac{ds}{dt}\right)^2}.$$

Integrating, observing that u is the initial velocity, and

$$2\kappa s = -\log \frac{g + \kappa\left(\frac{ds}{dt}\right)^2}{g + \kappa u^2};$$

$$\therefore \frac{ds}{dt} = v = u^2 e^{-2\kappa s} - \frac{g}{\kappa}\left(1 - e^{-2\kappa s}\right). \qquad (g)$$

At the highest point $v = 0$, which in (f) and (g) gives

$$t = (\kappa g)^{-\frac{1}{2}} \tan^{-1} u \left(\frac{\kappa}{g}\right)^{\frac{1}{2}}; \qquad (h)$$

and, $$s = \frac{1}{2\kappa} \log\left(1 + \frac{\kappa}{g} u^2\right). \qquad (i)$$

Substitute this value of s in equation (c) of the preceding example, and we have

$$v = \sqrt{\frac{g}{\kappa}\left(1 - \frac{1}{\log(1 + \frac{\kappa}{g} u^2)}\right)}$$

$$= \sqrt{\frac{g}{\kappa}\left(1 - \frac{1}{1 + \frac{\kappa}{g} u}\right)}$$

$$= \sqrt{\frac{g}{g + \kappa u^2}} \, u;$$

which gives the velocity in descending to the point from which it started; and as it is less than u, the velocity of return will be less than that with which it was thrown upward. This is because the resistance of the air is against the velocity during the entire movement, both upwards and downwards.

The same value of s (Eq. (i)) substituted in (d) of the preceding example gives the time of descent,

$$t = \frac{1}{2\sqrt{\kappa g}} \log \frac{\sqrt{g + \kappa u^2} + u\sqrt{\kappa}}{\sqrt{g \times \kappa u^2} - u\sqrt{\kappa}};$$

which differs from the time of the ascent, as given by (h) above.

31*st. Suppose that the force is attractive and varies inversely as the cube of the distance, and that the medium resists as the square of the velocity, and as the square of the density, the density varying inversely as the distance from the origin.*

Let $\kappa = $ the coefficient of resistance, being the resistance for a unit of density of the medium and a unit of velocity;

then $\frac{\kappa}{s^2}\left(\frac{ds}{dt}\right)^2$ = the resistance at any point.

$$\therefore \frac{d^2s}{dt^2} = -\frac{\mu}{s^3} + \frac{\kappa}{s^2}\left(\frac{ds}{dt}\right)^2.$$

Multiply by $2ds$, and we have

$$d\left(\frac{ds}{dt}\right)^2 - \frac{2\kappa}{s^2}\left(\frac{ds}{dt}\right)^2 ds = -\frac{2\mu}{s^3}ds.$$

This is a linear differential equation of which the integrating factor is $e^{\frac{2\kappa}{s}}$.

The initial values are $t = 0$, and $s = a$ for $v = u$;

$$\therefore e^{\frac{2\kappa}{s}}\left(\frac{ds}{dt}\right)^2 - e^{\frac{2\kappa}{a}}u^2 = 2\frac{\mu}{\kappa^2}\left\{\frac{2\kappa - s}{s}e^{\frac{2\kappa}{s}} - \frac{2\kappa - a}{a}e^{\frac{2\kappa}{a}}\right\},$$

which gives the velocity in terms of the space. The final integral cannot be found.

25. WORK AND VIS VIVA (*or living force.*)—Resuming equation (21), and multiplying both members by ds, we have

$$Fds = M\frac{d^2s}{dt^2}ds.$$

Integrating between the limits, $v = v_0$ for $s = 0$; and $v = v$ for $s = s$, we have

$$\int Fds = \tfrac{1}{2}M(v^2 - v_0^2). \qquad (23)$$

If $v_0 = 0$, we have

$$\int Fds = \tfrac{1}{2}Mv^2. \qquad (24)$$

The expression $\tfrac{1}{2}Mv^2$ is called the VIS VIVA (or *living force*) of a body whose mass is M and velocity v. Its physical importance is determined from the first member of the equation, which is called the work done by a force F in the space s. Hence *the vis viva equals the work done by the moving force.*

WORK, *mechanically, is overcoming resistance.* It requires a certain amount of work to raise one pound one foot, and twice that amount to raise two pounds one foot, or one pound two feet. Similarly, if it requires 100 pounds to move a load on a horizontal plane, a certain amount of work will be accomplished in moving it one foot, twice that amount in moving it two feet, and so on. Hence, generally, if

$F =$ a constant force which overcomes a constant resistance, and

$s =$ the space over which F acts projected on the action-line of the force, then

$$Work = U = Fs \, ; \qquad (25)$$

and similarly, if

$F =$ a *variable force*, then
$$Work = U = \Sigma F ds \, ; \qquad (26)$$

and if F is a function of s we have

$$U = \int F ds.$$

The UNIT of work is one pound raised vertically one foot.

The total work, according to equation (25), is independent of the time, since the space may be accomplished in a longer or shorter time.

But implicitly it is a function of the *time* and *velocity*. If the work be done at a uniform rate, we have

$$s = vt,$$

and
$$Fs = Fvt.$$

If $t = 1$, we have

$$Fv, \qquad (28)$$

which is called the *Dynamic Effect*, or *Mechanical Power*.

MECHANICAL POWER *is the rate of doing work*. It is measured by the amount of work done, or which the agent is capable of doing, in a unit of time when working uniformly. The *unit* most commonly employed is called the *horse-power*, which equals 33,000 pounds raised one foot per minute.

<small>Every moving body on the surface of the earth does work, for it overcomes a resistance, whether it be friction or resistance of the air, or some other resistance. The same is true of every body in the universe, unless it moves</small>

in a non-resisting medium.* Animals work not only as beasts of burden, but in their sports and efforts to maintain life; water as it courses the stream wears its banks or the bed, or turns machinery; wind fills the sail and drives the vessel, or turns the windmill, or in the fury of the tornado levels the forest, and often destroys the *works* of man. The raising of water into the air by means of evaporation; the wearing down of hills and mountains by the operations of nature; the destruction which follows the lightning-stroke, etc., are examples of work.

Work may be useful or prejudicial. That work is useful which is directly instrumental in producing useful effects, and prejudicial when it wears the machinery which produces it. Thus in drawing a train of cars, the useful work is performed in moving the train, but the prejudicial work is overcoming the friction of the axles, the friction on the track, the resistance of the air, the resistance of gravity on up grades, etc. It is not always possible to draw a practical line between the useful and prejudicial works, but the sum of the two always equals the total work done, and hence for economy the latter should be reduced as much as possible.

In order to determine practically the work done, the intensity of the force and the space over which it acts must be measured simultaneously. Some form of spring balance is commonly used to measure the force, and when thus employed is called a Dynamometer. It is placed between the moving force and the resistance, and the reading may be observed, or autographically registered by means of suitable mechanism. The corresponding space may also be measured directly, or secured automatically. There are many devices for securing these ends, and not a few make both records automatically and simultaneously.

If the force is not a continuous function of the space, equation (26) must be used. The result may be shown graphically by laying off on the abscissa, AB, the distances ac, ce, etc., proportional to the spaces, and erecting ordinates ab, cd, ef, etc., proportional to the corresponding forces, and joining their upper ends by a broken line, or, what is better, by a line which

---

* All space is filled with something, since light is transmitted from all directions. But is it not possible that there may be a *something* through which *bodies* may move without resistance?

is slightly curved, the amount and direction of curvature being *indicated* by the broken line previously constructed; and the area thus inclosed will *represent* the work. The area will be given by the formula
$\Sigma F . \overline{\Delta s}$.

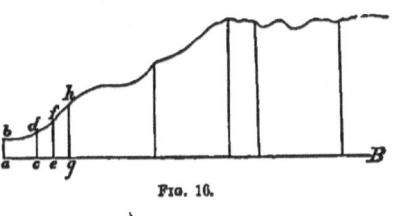

Fig. 16.

Simpson's rule for determining the area is:—

Divide the abscissa AB into an even number of equal parts, erect ordinates at the points of division, and number them in the order of the natural numbers. Add together four times the even ordinates, twice the odd ordinates and the extreme ordinates, and multiply the sum by one third of the distance between any two consecutive ordinates.

If y_0, y_1, y_2, etc., are the successive ordinates, and l the distance between any two consecutive ones, the rule is expressed algebraically as follows:—

$$Area = \tfrac{1}{3}l\,(y_0 + 2y_1 + 4y_2 + 2y_3 + 4y_4 + 2y_5 + \cdots y_n) \quad (29)$$

If the applied pressure, F, is exerted against a body which is perfectly free to move, generating a velocity v, then the work which has been expended is, equation (24), $\tfrac{1}{2}Mv^2$. This is called *stored work*, and the amount of work which will be done by the moving body in being brought to rest will be the same amount. If the body is not perfectly free the quantity $\tfrac{1}{2}Mv^2$ is the quantity of work which has been expended by so much of the applied force as exceeds that which is necessary in overcoming the frictional resistance. Thus a locomotive starts a train from rest, and when the velocity is small the power exerted by the locomotive may exceed considerably the resistances of friction, air, etc., and produce an increasing velocity, until the resistances equal constantly the tractive force of the locomotive, after which the velocity will be uniform. The work done by the locomotive in producing the velocity v in excess of that done in overcoming the resistances will be $\tfrac{1}{2}Mv^2$, in which M is the mass of the train, including the locomotive.

We see that double the velocity produces four times the work. This is because twice the force produces twice the velocity, and hence the body will pass over twice the space in the same time, so that in producing double the velocity we have $2F.2s = 4Fs$, and similarly for other velocities.

[We have no single word to express the unit of living force. If a unit of mass moving with a velocity of one foot per second be the unit of living force, and be called a *Dynam*, then would the living force for any velocity and mass be a certain number of *Dynams*.]

Since *work is not force*, but the effect of a force exerted through a certain space, independently of the time, we call it, for the sake of brevity, *space-effect*.

Vis viva, or living force, is not force, but it is twice the work stored in a moving mass. It equals twice the *space-effect*.

[The expression Mv^2 was called the vis viva in the first edition of this work, and is still so defined by many writers; but there appears to be a growing tendency towards the general adoption of the definition given in the text. It is immaterial which is used, provided it is always used in the same sense.]

Examples.

1. A body whose weight is 10 pounds is moving with a velocity of 25 feet per second; required the amount of work which will be done in bringing it to rest.

Ans. 97.2 foot-pounds.

2. A body falls by the force of gravity through a height of h feet; required the work stored in it.

Let $W =$ the weight of the body,
$M =$ the mass of the body,
$g =$ acceleration due to gravity, and
$v =$ the final velocity, then
$v^2 = 2gh$, and $Mg = W$;

$$\therefore \tfrac{1}{2}Mv^2 = \frac{W}{2g} \cdot 2gh = Wh.$$

3. A body whose weight is 100 pounds is moving on a horizontal plane with a velocity of 15 feet per second; how far will it go before it is brought to rest, if the friction is constantly 10 lbs?

$Ans. = 34.6 +$ ft.

4. A hammer whose weight is 2000 pounds has a velocity of 20 feet per second; how far will it drive a pile if the constant resistance is 10,000 pounds, supposing that the whole *vis viva* is expended in driving the pile?

5. If a train of cars whose weight is 100,000 pounds is moving with a velocity of 40 miles per hour, how far will it move before it is brought to rest by the force of friction, the friction being 8 pounds per ton, or $\frac{8}{2000}$ of the total weight?

6. If a train of cars weighs 300 tons, and the frictional resistance to its movement is 8 pounds per ton; required the horse-power which is necessary to overcome this resistance at the rate of 40 miles per hour.

$Ans.$ 256.

7. If the area of a steam piston is 75 square inches, and the steam pressure is 60 pounds per square inch, and the velocity of the piston is 200 feet per minute, required the horse-power developed by the steam.

8. If a stream of water passes over a dam and falls through a vertical height of 16 feet, and the transverse section of the stream at the foot of the fall is one square foot, required the horse-power that is constantly developed. $Ans.$ 58.2 +.

Let $g = 32\frac{1}{6}$ feet, and the weight of a cubic foot of water, $62\frac{1}{2}$ lbs.

$Ans.$ 21.89.

9. A steam hammer falls vertically through a height of 3 feet under the action of its own weight and a steam pressure of 1000 pounds. If the weight of the hammer is 500 pounds, required the amount of work which it can do at the end of the fall.

26. ENERGY *is the capacity of an agent for doing work.* The energy of a moving body is called *actual* or *Kinetic energy*, and is expressed by $\frac{1}{2}Mv^2$. But bodies not in motion may have a capacity for work when the restraining forces are removed. Thus a spring under strain, water stored in a mill-dam, steam in a boiler, bodies supported at an elevation, etc., are examples of stored work which is *latent*. This is called *Potential energy*. A *moving* body may possess potential energy entirely distinct from the actual. Thus, a locomotive boiler containing steam, may be moved on a track, and the kinetic energy would be expressed by $\tfrac{1}{2}Mv^2$, in which M is the mass of the boiler, but the *potential* energy would be the amount of work which the steam is capable of doing when used to run machinery, or is otherwise *employed*. These principles have been generalized into a law called the *Conservation of energy*, which implies that the total energy, including both Kinetic and Potential, in the universe remains constant. It is made the fundamental theorem of modern physical science.

The energy stored in a moving body is not changed by changing the direction of its path, *provided* the velocity is not changed; for its energy will be constantly expressed by $\tfrac{1}{2}Mv^2$. Such a change may be secured by a force acting continually normal to the path of the moving body; and hence we say *that a force which acts continually perpendicular to the path of a moving body does no work upon the body*. Thus, if a body is secured to a point by a cord so that it is compelled to move in the circumference of a circle; the tension of the string does no work, and the vis viva is not affected by the body being constantly deflected from a rectilinear path.

MOMENTUM.

27. Resuming again equation (21), multiplying by dt, and integrating gives,

$$\int_0^t F dt = M \int \frac{d^2s}{dt} = M \frac{ds}{dt} = Mv. \qquad (30)$$

The expression Mv is called *momentum*, and by comparing it with the first member of the equation we see that it is the effect of the force F acting during the time t, and is independent of the space. For the sake of brevity we may call the momentum a *time-effect*.

If the body has an initial velocity we have

$$\int_{t_0}^{t} F dt = M(v - v_0); \qquad (31)$$

which is the momentum gained or lost in passing from a velocity v_0 to v.

Momentum is sometimes called *quantity of motion*, on account of its analogy to some other quantities. Thus the intensity of heat depends upon temperature, and is measured in degrees; but the quantity of heat depends upon the volume of the body containing the heat and its intensity. The intensity of light may be uniform over a given surface, and will be measured by the light on a unit of surface; but the quantity is the product of the area multiplied by the intensity. The intensity of gravity is measured by the acceleration which is produced in a falling body, and is independent of the mass of the body; but the quantity of gravity (or total force) is the product of the mass by the intensity (or Mg). Similarly with momentum. The *velocity* represents the *intensity* of the motion, and is independent of the mass of the body; but the quantity of motion is the product of the mass multiplied by the velocity.

Differentiating (30) and reducing, gives

$$F = M \frac{dv}{dt};$$

which is the same as (18), and in which $\frac{dv}{dt}$ is a velocity-increment; *hence the momentum impressed each instant is a measure of the moving force.*

If the force F is constant we have from (30),

$$Ft = Mv;$$

and for another force F' acting during the same time

$$F't = M'v';$$

$$\therefore F : F' :: Mv : M'v';$$

hence, the forces are directly as the momenta produced by them respectively.

If the forces are variable, let

$$\int_0^t F dt = Q = Mv, \text{ and } \int_0^{t'} F' dt = Q' = M'v';$$

then
$$Q : Q' :: Mv : M'v';$$

hence the *time-effects* are directly as the momenta impressed.

We thus have several distinct quantities growing out of equation (21) of which the English units are as follows :—

The unit of force, F, is.......................... 1 ℔.
The unit of work or space effect is................ 1 ℔ × 1 ft.
The unit of vis viva is.............. 1 ℔ of mass × 1² ft. × 1 sec.
The unit of momentum............. 1 ℔ of mass × 1 ft. × 1 sec.

IMPULSE.

28. *An impulse* is the *effect of a blow.* When one body strikes another, an impact is said to take place, and certain effects are produced upon the bodies. These *effects* are produced in an exceedingly short time, and for this reason they are sometimes called *instantaneous forces;* which, being strictly defined, means a force which produces its effect *instantly, requiring no time for its action;* but no such force exists in nature. The law of action during impact is not generally known, but it must be some function of the time.

Resuming equation (31), we have

$$\int_{t_0}^t F dt = M(v - v_0);$$

which is true, whatever be the relation between the force F and the time t. If the initial velocity of the body be zero, we have

$$v_0 = 0,$$

and
$$\int_{t_0}^t F dt = MV = Q.$$

We see from the above equation that as t diminishes F must increase to produce the same *effect.* We see that in this case the first member is the *time-effect* of an impulse, and the second member measures its effect in producing a change of velocity. Calling this value Q, we have

$$Q = M(v - v_0) = MV. \qquad (31a)$$

Hence, the measure of an impulse in producing a change of velocity of a body is the increased (or decreased) momentum produced in the body.

This is the same as when the force and time are finite. If the force were *strictly* instantaneous the velocity would be changed from v_0 to v without moving the body, since it would have no time in which to move it.

Similarly from equation (23) we have

$$\int_0^s F ds = \tfrac{1}{2} M(v^2 - v_0^2);$$

in which for an impulse F will be indefinitely large; and hence *the work done by an impulse is measured in the same way as for finite forces.*

All the *effects* therefore of an impulse are measured in the same way as the *total effects* produced by a finite force.

In regard to *forces*, we investigate their laws of action; or having those laws and the initial condition of the body we may determine the velocity, energy, or position of the body at any instant of time or at any point in space, and hence we may determine final results; but in regard to *impulses* we determine only certain final results without assuming to know anything of the laws of action of the forces, or of the time or space occupied in producing the effect.

The terms "*Impulsive force,*" and "*Instantaneous force,*" are frequently used to denote the effect of an *Impact;* but since the effect is not a force, they are ambiguous, and the term *Impulse* appears to be more appropriate.

An *incessant force* may be considered as the action of an infinite number of infinitesimal impulses in a finite time.

The question is sometimes asked, " What is the force of a

blow of a hammer?" If by *the force* is meant the pressure in pounds between the face of the hammer and the object struck, it cannot be determined unless the law of resistance to compression between the bodies is known during the contact of the bodies. But this law is generally unknown. The pressure begins with nothing at the instant of contact and increases very rapidly up to the instant of greatest compression, after which the pressure diminishes. The pressure involves the elasticity of both bodies; the rapidity with which the force is transmitted from one particle to another; the amount of the distortion; the pliability of the bodies; the duration of the impact; and some of these depend upon the degree of fixedness of the body struck; and several other minor conditions; and hence we consider it impossible to tell *exactly* what the *force* is.

Examples.

1. Two bodies whose weights are W and W_1 are placed very near each other, and an explosive is discharged between them; required the relative velocities after the discharge.

2. A man stands upon a rough board which is on a perfectly smooth plane, and jumps off from the board; required the relative velocities of the man and board.

[Obs. The common centre of gravity of the man and board will remain the same after they separate that it was before. After separating they would move on forever if they did not meet with any obstacle to prevent their motion.]

3. A man whose weight is 150 pounds walks from one end of a rough board to the other, which is twelve feet long, and free to slide on a perfectly smooth plane; if the board weighs 50 pounds, required the distance travelled by the man in space.

4. In example 3 of article 24, suppose that the weight 10 pounds is permitted to fall freely through a height h, when it produces an impulse on the body (50 pounds) through the intermediate inextensible string; required the initial velocity of the body.

Let $v_0 = \sqrt{2gh}$ = the velocity of the weight just before the impulse; and

v = the velocity immediately afterward, which will be the common velocity of the body and weight;

then
$$Q = \frac{50}{g} v = \frac{10}{g} (v_0 - v) ;$$
$$\therefore v = \tfrac{1}{6} v_0 .$$

The subsequent motion may be found by equation (21), observing that the initial velocity is v.

The tension on the string will be infinite if it is inextensible, but practically it will be finite, for it will be more or less elastic.

[Some writers have used the expression *impulsive tension* of the string instead of *momentum*.]

5. If a shell is moving in a straight line, in vacuo, with a velocity v, and bursts, dividing into two parts, one part moving directly in advance with double the velocity of the body; what must be the ratio of the weights of the two parts so that the other part will be at rest after the body bursts?

6. Explain how a person sitting in a chair may move across a room by a series of jerks without touching the floor. (Can he advance if the floor is perfectly smooth?)

7. A person is placed on a perfectly smooth plane, show how he can get off if he cannot reach the edge of the plane.

The same impulse applied to a small body will impart a greater amount of energy than if applied to a large one. Thus, in the discharge of a gun, the impulse imparted to the gun equals that imparted to the ball, but the work, or destructive effect, of the gun is small compared with that of the ball. The *time of the action* of the explosive is the same upon both bodies, but the space moved over by the gun will be small compared with that of the ball during that time.

The product Mv, being the same for both, as M decreases v increases, but the work varies as the square of the velocity.

DIRECT CENTRAL IMPACT.

29. If two bodies impinge upon one another, so that the line of motion before impact passes through the centre of the bodies, it is said to be *central;* and if at the same time the common tangent at the point of contact is perpendicular to the line of motion, it is said to be *direct* and *central.* If their common tangent is perpendicular to the line of motion, but if the latter does not pass through the centre of the body impinged upon, it is called *eccentric impact.* In this place, we consider only the simplest case; that in which the impact is *direct* and *central*.

When two bodies impinge directly against one another, whether moving in the same or opposite directions, they mutually displace the particles in the vicinity of the point of contact, producing compression which goes on increasing until it becomes a maximum, at which instant they have a common velocity. A complete analysis of the motion during contact involves a knowledge of the motion of all the particles of the mass, and would make an exceedingly complicated problem, but the motion at the instant of maximum compression may be easily found if we assume that the compression is instantly distributed throughout the mass.

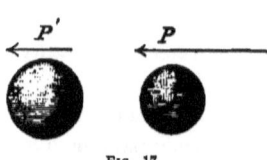

Fig. 17.

Let M_1 and M_2 be the respective masses of the bodies;

v_1 and v_2 the respective velocities before impact;

v_1' and v_2' the respective velocities at the instant of maximum compression, and

Q_1 and Q_2 the momenta lost respectively by the bodies during compression.

Then from (31)

$$Q_1 = M_1(v_1 - v_1');$$

which is the momentum lost by M_1 on account of the action of M_2. Similarly

$$Q_2 = M_2(v_2 - v_2');$$

which will be essentially negative if the bodies move in the same direction, and will be the momentum gained by M_2 on account of the action of M_1.

But at the instant of greatest compression

$$v_1' = v_2';$$

and, because they are in mutual contact during the same time, their *time-effects* are equal, but in opposite directions,

$$\therefore Q_1 = -Q_2.$$

Combining these four equations, we find by elimination

$$Q_1 = \frac{M_1 M_2}{M_1 + M_2}(v_1 - v_2) = -Q_2; \qquad (32)$$

$$v_1' = \frac{M_1 v_1 + M_2 v}{M_1 + M_2} = v_2'; \qquad (33)$$

which velocity remains constant for perfectly non-elastic bodies after impact, since such bodies have no power of restitution and will move on with a common velocity.

DIRECT CENTRAL IMPACT OF ELASTIC BODIES.

30. ELASTIC BODIES are such as regain a part or all of their distortion when the distorting force is removed. If they regain their original form they are called *perfectly elastic*, but if only a part, they are called *imperfectly elastic*. After the impact has produced a maximum compression, the elastic force of the bodies causes them to separate, but all the effect which the force of restitution can produce upon the movement of the bodies, evidently takes place while they are in contact. If they are perfectly elastic and do not fully regain their form at the instant of separation, they will continue to regain their form after separation, but the latter effect we do not consider in this place. The ratio between the forces of compression and those of restitution has often been called the *modulus of elasticity*, but as some ambiguity results from this definition, we will call it the *modulus of restitution*. At every point of the restitution there is assumed to be a constant ratio between the force due to compression and that to restitution. But it is unnecessary for present purposes to *trace* these effects, for by equation (31) we may determine the result when the bodies finally separate from each other.

Let $e_1 =$ the ratio of the force of compression to that of restitution of one body, which is called the *modulus of restitution*.

$e_2 =$ the corresponding value for the other;

$V_1 =$ the velocity of M_1 at the instant when they separate from each other; and

$V_2 =$ the corresponding velocity for M_2.

Then from equation (31)

$$e_1 Q_1 = M_1 (V_1 - v_1'); \qquad (34)$$

$$e_2 Q_2 = M_2 (V_2 - v_2'). \qquad (35)$$

As before $Q_1 = - Q_2$ and we will also assume that $e_1 = e_2 = e$. These combined with (32) and (33) give

$$V_1 = \frac{M_1 v_1 + M_2 v_2}{M_1 + M_2} - \frac{e M_2}{M_1 + M_2}(v_1 - v_2); \quad (36)$$

$$V_2 = \frac{M_1 v_1 + M_2 v_2}{M_1 + M_2} + \frac{e M_1}{M_1 + M_2}(v_1 - v_2). \quad (37)$$

31. DISCUSSION OF EQUATIONS (36) and (37).

1°. If the bodies are perfectly non-elastic, $e = 0$.

$$\therefore V_1 = \frac{M_1 v_1 + M_2 v_2}{M_1 + M_2} = V_2; \qquad (38)$$

which is the same as (33).

2°. If the restitution is perfect $e = 1$.

$$\therefore V_1 = v_1 - \frac{2 M_2}{M_1 + M_2}(v_1 - v_2); \qquad (39)$$

$$V_2 = v_2 + \frac{2 M_1}{M_1 + M_2}(v_1 - v_2). \qquad (40)$$

[31.] IMPACT. 59

From (38) we have

$$V_1 - v_1 = -\frac{M_2}{M_1 + M_2}(v_1 - v_2);$$

and,
$$V_2 - v_2 = \frac{M_1}{M_1 + M_2}(v_1 - v_2).$$

Similarly from (39) and (40)

$$V_1 - v_1 = -\frac{2M_2}{M_1 + M_2}(v_1 - v_2);$$

$$V_2 - v_2 = \frac{2M_1}{M_1 + M_2}(v_1 - v_2);$$

hence, the velocity lost by one body and gained by the other is twice as much when the bodies are perfectly elastic as when they are perfectly non-elastic.

3°. If $M_1 = M_2$, then for perfect restitution we have

$$V_1 = v_1 - \frac{2M_1}{M_1 + M_1}(v_1 - v_2) = v_2;$$

$$V_2 = v_2 + \frac{2M_1}{M_1 + M_1}(v_1 - v_2) = v_1;$$

that is, they will interchange velocities.

4°. If M_1 impinges against a fixed body, we have $M_2 = \infty$, and $v_2 = 0$.

$$\therefore V_1 = -e\, v_1.$$

This furnishes a convenient mode of determining e. For if a body falls from a height h upon a fixed horizontal plane, it will rebound to a height h_1;

$$\therefore h_1 = e^2 h, \text{ or } e = \sqrt{\frac{h_1}{h}}.$$

Also if $e = 1$
$$V_1 = -v_1;$$

or the velocity after impact will be the same as before, but in an opposite direction.

Also if $e = 0$, $V_1 = 0$; or the velocity will be destroyed.

5°. If $v_2 = 0$ we have

$$V_1 = \frac{M_1 - eM_2}{M_1 + M_2} v_1;$$

$$V_2 = \frac{M_1 + eM_1}{M_1 + M_2} v_1.$$

$\qquad$ (41)

EXAMPLES.

(1.) A mass M_1 with a velocity of 10, impinges on M_2 moving in an opposite direction, moving with a velocity 4 and has its velocity reduced to 5; required the relative magnitudes of M_1 and M_2.

(2.) Two inelastic bodies, weighing 8 and 5 pounds respectively, move in the same direction with velocities 7 and 3; required the common velocity after impact, and the velocity lost and gained by each.

(3.) If M_1 weighs 12 pounds and moves with a velocity of 15, and is impinged upon by a body M_2 weighing 16 pounds, producing a common velocity of 30, required the velocity of M_2 before impact if it moves in the same or opposite direction.

(4.) If $5M_1 = 6M_2$, $6v_1 = -5v_2$, $v_2 = 7$, and $e = \frac{2}{3}$; required the velocity of each after impact.

(5.) If $M_1 = 2M_2$. $V_1 = \frac{2}{3}v_1$, and $v_2 = 0$; required e.

(6.) If v_1 is 26, M_2 is moving in an opposite direction with a velocity of 16; $M_1 = 2M_2$, $e = \frac{2}{3}$; required the distance between them $5\frac{1}{4}$ seconds after impact.

(7.) Two bodies are perfectly elastic and move in opposite directions; the weight of M_1 is twice M_2, but $v_2 = 2v_1$; required the velocities after impact.

(8.) There is a row of perfectly elastic bodies in geometrical progression whose common ratio is 3, the first impinges on the second, the second on the third and so on; the last moves off with $\frac{1}{64}$ the velocity of the first. What is the number of bodies?

Ans. 7.

LOSS OF VIS VIVA IN THE IMPACT OF BODIES.

32. Before impact the vis viva of both bodies was

$$M_1 v_1^2 + M_2 v_2^2;$$

and after impact

$$M_1 V_1^2 + M_2 V_2^2;$$

which by means of (36) and (37) becomes

$$M_1 V_1^2 + M_2 V_2^2 = M_1 \dot{v}_1^2 + M_2 v_2^2 - \frac{(1-e^2)}{M_1+M_2} \frac{M_1 M_2}{}(v_1-v_2)^2. \quad (42)$$

For perfectly elastic bodies $e = 1$ and the last term disappears; hence *in the impact of perfectly elastic bodies no vis viva is lost.*

If the bodies are imperfectly elastic e is less than 1, and since $(v_1 - v_2)^2$ is always positive, it follows that *in the impact of imperfectly elastic bodies vis viva is always lost, and the greatest loss is suffered when the bodies are perfectly non-elastic.*

If $e = 0$, (42) becomes

$$M_1(v_1^2 - V_1^2) + M_2(v_2^2 - V_2^2) = \frac{M_1 M_2}{M_1 + M_2}(v_1 - v_2)^2 ; \quad (43)$$

in which each member is the total loss by both bodies. It is also the loss up to the instant of greatest compression when the bodies are elastic.

If M_2 is very large compared with M_1 we have from (38)

$$V_1 = v_2 \text{ nearly}, = V_2,$$

and (43) becomes

$$M_1 v_1^2 - M_1 V_1^2 = M_1 (v_1 - V_1)^2,$$

the second member of which is frequently used in hydraulics for finding the *vis viva* lost by a sudden change of velocity.

THESE INVESTIGATIONS show the great utility of springs in vehicles and machines which are subjected to impact.

RELATIONS OF FORCE, MOMENTUM, WORK, AND VIS VIVA.

33. WE MAY NOW DETERMINE THE EXACT OFFICE in the same problem of the quantities ;—*force, momentum, work, and vis viva.* Suppose that a force, whether variable or constant, impels a body, it will in a time t generate in the mass M a certain velocity v. This *force* may at any instant of its action be measured by a certain number of pounds or its equivalent

Suppose that this mass impinges upon another body, which may be at rest or in motion. In order to determine the effect upon their *velocities* we use the principle of *momentum*, as has been shown. But the bodies are compressed during impact and hence *work* is done. The amount of work which they are capable of doing is equal to the sum of their *vis viva*; and if they are brought to rest all this work is expended in compressing them. If the velocity of a body after impact is less than that before, it has done an amount of work represented by $\frac{1}{2}M(v^2 - V^2)$, and similarly if the other body has its velocity increased kinetic energy is imparted to it. *The distortions of bodies represent a certain amount of work expended.* And this explains why in the impact of imperfectly elastic bodies *vis viva* is always lost, for a portion of the distortion remains. But no force is lost. One of the grandest generalizations of physical science is, *that no force in nature is lost.* In the case of impact, compression develops heat, and this passes into the air or surrounding objects, and the amount of energy which is stored in the heat, electricity or other element or elements, which is developed by the compression, exactly equals that lost to the masses. We thus see that in the case of moving bodies, *force impels, momentum determines velocity after impact, and work or vis viva represents the resistance which the particles offer to being displaced.*

34. STATICS is that case in which the force or forces which would produce motion are instantly arrested, resulting in pressure only. The expression for the elementary work which a force can do is Fds, but if the space vanishes, we have, $Fds = 0$. This, as we shall see hereafter, is a special case of "virtual velocities."

The forces which act upon a body may be in equilibrium and yet motion exist, but in such cases the velocity is uniform.

35. The term *power* is often used in the same sense as *force*, but generally it refers to an acting agent. The term *mechanical power* is not only recognized in this science, but has a specific meaning, and for the purpose of avoiding ambiguity, it is better to use the term *effort* in reference to mechanical agents. Thus, instead of saying *the power and weight*, as is often done, say *the effort and resistance*.

36. INERTIA *implies passiveness or want of power.* It means that matter has no power within itself to put itself in motion, or when in motion to change its rate of motion. Unless an external force be applied to it, it would, if at rest, remain forever in that condition; or if in motion, continue forever in motion. Gravity, which is a force apparently inherent in matter, can produce motion only by its action upon other matter.

INERTIA *is not a force*, but because of the property above explained, those impressed forces which produce motion are measured by the product of the mass into the acceleration as explained in preceding articles; and many writers call this MEASURE *the force of inertia.*

37. NEWTON'S THREE LAWS OF MOTION.

Sir Isaac Newton expressed the fundamental principles of motion in the form of three laws or mechanical axioms; as follows:—

1st. Every body continues in its state of rest or of uniform motion in a straight line unless acted upon by some external force.

2d. Change of motion is proportional to the force impressed, and is in the *direction* of the line in which the force acts.

3d. To every action there is opposed an equal reaction.

[As simple as these laws appear to the student of the present day, the science of Mechanics made no essential progress until they were recognized. See Whewell's *Inductive Sciences*, 3d ed., vol. 1, p. 311.]

38. In all the problems thus far considered, it has been assumed that the action-line of the force or forces passed through the centre of the mass, producing a motion of translation only. But if the action-line does not pass through the centre, it will produce both translation and rotation.

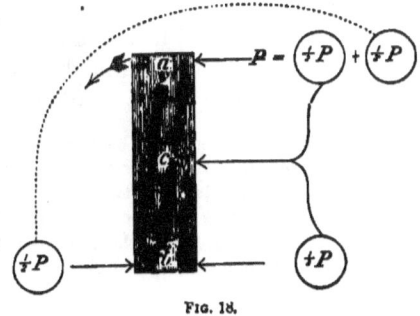

FIG. 18.

In eccentric impact both translation and rotation is produced. The centre of the body will move in a straight line, but every other point will describe arcs of circles in reference to the centre of the body, which in space will be curves more or less elongated. The velocity of translation will be directly proportional to the intensity of the impulse imparted to the body, but the angular velocity will depend upon the intensity of the impulse and the distance of the point of impact from the centre of the body.

In Figure 18, let $Q = Mv$ be the impulse imparted to the body; in which M is the mass of the body and v the velocity of the centre. Let this impulse be imparted at a. At b, a distance from the centre $= cb = ac$, let two equal and opposite impulses be imparted, each equal to $\tfrac{1}{2}Q$. The impulse Q, equals $\tfrac{1}{2}Q + \tfrac{1}{2}Q$. The four impulses evidently produce the same effect upon the body as the single impulse Q. If now one of the impulses, $\tfrac{1}{2}Q$, above the centre is combined with the equal and parallel one acting in the same direction below the centre, their effect will be equivalent to a single one, equal to Q applied at the centre c. This produces translation only. The other $\tfrac{1}{2}Q$ above the centre combined with the equal and opposite $\tfrac{1}{2}Q$ below the centre, produces rotation only; and it is evident that the greater the distance a, the point of impact, is from the centre, the greater will be the amount of rotation.

An impact (or blow) at a to produce a velocity v at the centre of the body, must act through a greater space during contact, or the impacting body must move with a greater velocity, than if the impact be in a line passing through the centre c. Such an impulse at a will impart more *energy* to the body than the one at c; for there will be the same energy due to translation in both cases, viz., $\tfrac{1}{2}Mv^2$, and in the former case there will be an additional amount due to rotation.

(The entire energy stored in the body will be $\tfrac{1}{2}Mv^2 + \tfrac{1}{2}I_m\omega^2$, in which I_m is the moment of inertia of the rotating mass in reference to an axis through the centre, and ω is the angular velocity in reference to the same axis; and the other notation is the same as in the preceding Article. See Article 127. This expression for the energy, in case the bodies are perfectly elastic, will equal the energy lost by the impacting body.)

CHAPTER II.

COMPOSITION AND RESOLUTION OF FORCES.

CONCURRENT FORCES.

39. If two or more forces act upon a material particle, they are said to be *concurrent*. They may all act towards the particle, or from it, or some towards and others from.

40. If several forces act along a material line, they are called *conspiring* forces, and their effect will be the same as if all were applied at the same point.

41. The Resultant of two or more concurrent forces is that force which if substituted for the system will produce the same effect upon a particle as the system.

Therefore, if a force equal in magnitude to the resultant and acting along the same action-line, but in the opposite direction, be applied to the same particle, the system will be in equilibrium.

If the resultant is negative, the equilibrating force will be positive, and vice versa.

Hence, if several concurrent forces are in equilibrium, any one may be considered as equal and opposite to the resultant of all the others.

42. *The resultant of several conspiring forces, equals the algebraic sum of the forces.* That is, if F_1, F_2, F_3, etc., are the forces acting along the same action-line, some of which may be positive and the others negative, and R is the resultant; then

$$R = F_1 + F_2 + F_3 + \text{etc.} = \Sigma F. \qquad (45)$$

43. *If two concurring forces be represented in magnitude and direction by the adjacent sides of a parallelogram, the resultant will be represented in magnitude and direction by the diagonal of the parallelogram.* This is called the *parallelogram of forces*.

If each force act upon a particle for an element of time it will generate a certain velocity. See equation (44). Let the velocity which F would produce be represented by AB; and that of P by $AD = BC$. These represent the spaces

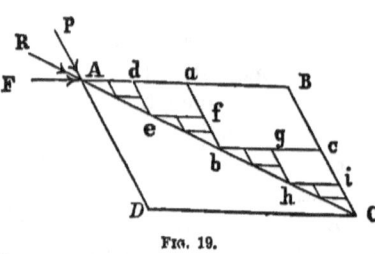

Fig. 19.

over which the forces respectively would move the particle in a unit of time if each acted separately. If we conceive that the force F moves it from A to B and that the motion is there arrested, and that P is then applied at B, but acting parallel to AD, then will the particle, at the end of two seconds, be at C. If, next, we conceive that each force acts alternately during one-half of a second beginning again at A, the particle will be found at a in one-half of a second; at b at the end of one second; at c at the end of one and one-half seconds; and finally at C at the end of two seconds. If the times be again subdivided the path will be Ad, de, ef, fb, bg, gh, hi, and iC, and it will arrive at C in the same time as before.

As the divisions of the time increase, the number of sides of the polygon increase, each side becoming shorter; and the polygonal path approaches the straight line as a limit. Therefore at the limit, when the force P and F act simultaneously, the particle will move along the diagonal, AC, of the parallelogram. But when they act simultaneously, they will produce their effect in the same time as each when acting separately; and hence, the particle will arrive at C at the end of *one second*. Therefore, a single force R, which is represented by AC, will produce the same effect as P and F, and will be the resultant. If now a force equal and opposite to R act at the same point as the forces F and P, the motion will be arrested and pressure only will be the result. See article 34. Hence, the parallelogram of velocities and of pressures becomes established.*

---

* This is one of the most important propositions in Mechanics, and has been proved in a variety of ways. One work gives forty-five different proofs. A demonstration given by M. Poisson is one of the most noted of the analytical proofs. Many persons object to admitting the idea of motion in proving the

If θ be the angle between the sides of the parallelogram which represent the forces P and F, and R be the diagonal, or resultant, we have from trigonometry

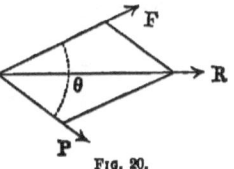
Fig. 20.

$$R^2 = F^2 + P^2 + 2PF \cos \theta. \quad (46)$$

If θ exceeds 90 degrees, it must be observed in the solution of problems that $\cos \theta$ will be negative.

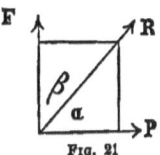
Fig. 21.

If $\theta = 90$ degrees, we have

$$R^2 = F^2 + P^2.$$

Also, if $\theta = 90°$, and a be the angle between R and P; and β between R and F; then

$$\left. \begin{array}{l} P = R \cos a \, ; \\ F = R \cos \beta = R \sin a. \end{array} \right\} \quad (47)$$

Squaring and adding, we have

$$P^2 + F^2 = R^2,$$

as before.

The forces P and F are called *component forces*, or simply *components*.

44. TRIANGLE OF FORCES. *If two forces are represented in magnitude and direction by two sides of a triangle taken in their order, the resultant will be represented in magnitude and direction by the third side.*

Thus, in Fig. 19, if AB and BC represent two forces in magnitude and direction, AC will represent the resultant.

---

parallelogram of pressures; but we have seen that a pressure when acting upon a free body will produce a certain amount of motion, and that this motion is a measure of the pressure, and hence its use in the proof appears to be admissible. But the strongest proof of the correctness of the proposition is the fact that in all the problems to which it has been applied, the results agree with those of experience and observation.

Since the sines of the angles of a triangle are proportional to the sides opposite, we have

$$\frac{F}{\sin \widehat{P,R}} = \frac{P}{\sin \widehat{F,R}} = \frac{R}{\sin \widehat{F,P}}. \qquad (48)$$

POLYGON OF FORCES.

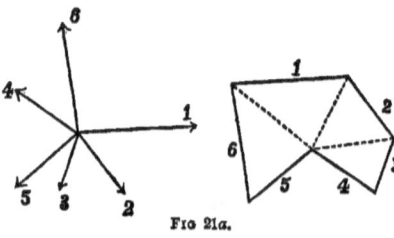

Fig. 21a.

45. *If several concurrent forces are represented in magnitude and direction by the sides of a closed polygon taken in their order, they will be in equilibrium.*

This may be proved by finding the resultant of two forces by means of the triangle of forces; then the resultant of that resultant and another force, and so on.

PARALLELOPIPED OF FORCES.

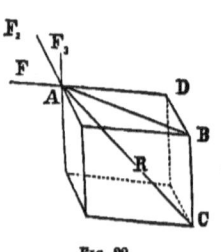

Fig. 22.

46. *If three concurrent forces not in the same plane are represented in magnitude and direction by the adjacent edges of a parallelopipedon, the resultant will be represented in magnitude and direction by the diagonal; and conversely if the diagonal of a parallelopipedon represents a force, it may be considered as the resultant of three forces represented by the adjacent edges of the parallelopipedon.*

In Fig. 22, if AD represents the force F_1 in magnitude and direction, and similarly DB represents F_2, and BC, F_3; then according to the triangle of forces AB will represent the resultant of F_1 and F_2; and AC the resultant of AB and F_3, and hence it represents the resultant of F_1, F_2, and F_3.

If F_1, F_2, and F_3 are at right angles with each other, we have

$$R^2 = F_1^2 + F_2^2 + F_3^2;$$

and if α is the angle $R\hat{,}F_1$, β of $R\hat{,}F_2$, and γ of $R\hat{,}F_3$; then

$$\left. \begin{array}{l} F_1 = R \cos \alpha; \\ F_2 = R \cos \beta; \\ F_3 = R \cos \gamma. \end{array} \right\} \qquad (49)$$

Squaring these and adding, we have

$$R^2 = F_1'^2 + F_2'^2 + F_3'^2, \text{ as before.}$$

EXAMPLES.

1. When $F = F_1$ and $\theta = 60°$, find R; (See Eq. (46)).
   Ans. $R = F\sqrt{3}$.

2. If $F = F_1$ and $\theta = 120°$, find R.

3. If $F = F_1$ and $\theta = 135°$, find R.
   Ans. $R = F\sqrt{2 - \sqrt{2}}$.

4. If $F = 2F_1 = 3R$, find θ.

5. If $\frac{1}{3}F = \frac{1}{4}F_1 = R$, find the angle $F\hat{,}F_1$.
   Ans. $90°$.

6. If $F = 7$, $F_1 = 9$, and $\theta = 25°$, find R and angle $F\hat{,}R$.

7. A cord is tied around a pin at a fixed point, and its two ends are drawn in different directions by forces F and P. Find θ when the pressure upon the pin is $R = \frac{1}{2}(P + F)$.

$$\text{Ans. } \cos \theta = \frac{2PF - 3(P^2 + F^2)}{8PF}.$$

8. When the concurring forces are in equilibrium, prove that

$$P : F : R :: \sin F\hat{,}R : \sin P\hat{,}R : \sin P\hat{,}F.$$

9. If two equal rafters support a weight W at their upper ends, required the compression on each. Let the length of

each rafter be a and the horizontal distance between their lower ends be b.

$$Ans. \quad \frac{a}{\sqrt{4a^2-b^2}} W.$$

10. If a block whose weight is 200 pounds is so situated that it receives a pressure from the wind of 25 pounds in a due easterly direction, and a pressure from water of 100 pounds in a due southerly direction; required the resultant pressure and the angle which the resultant makes with the vertical.

RESOLUTION OF CONCURRENT FORCES.

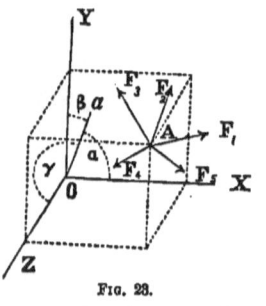

Fig. 23.

47. Let there be many concurrent forces acting upon a single particle, and the whole system be referred to rectangular co-ordinates.

Let F_1, F_2, F_3, etc., be the forces acting upon a particle at A;
x, y, z the co-ordinates of A;
a_1, a_2, etc., the angles which the direction-lines of the respective forces make with the axis of x;
β_1, β_2, etc., the angles which they make with y;
γ_1, γ_2, etc., the angles which they make with z; and
X, Y, and Z, the algebraic sum of the components of the forces when resolved parallel to the axes x, y, and z, respectively.

Then, according to equations (45) and (49), we have for equilibrium;

$$\left.\begin{aligned} X &= F_1 \cos a_1 + F_2 \cos a_2 + F_3 \cos a_3 + \text{etc.} = \Sigma F \cos a = 0\,; \\ Y &= F_1 \cos \beta_1 + F_2 \cos \beta_2 + F_3 \cos \beta_3 + \text{etc.} = \Sigma F \cos \beta = 0\,; \\ Z &= F_1 \cos \gamma_1 + F_2 \cos \gamma_2 + F_3 \cos \gamma_3 + \text{etc.} = \Sigma F \cos \gamma = 0\,; \end{aligned}\right\} \quad (50)$$

If they are not in equilibrium, let R be the resultant, and by introducing a force equal and opposite to the resultant, the system will be in equilibrium.

Let a, b and c be the angles which the resultant makes with the axes x, y and z respectively; then

$$X = R \cos a;$$
$$Y = R \cos b; \qquad (51)$$
$$Z = R \cos c.$$

Squaring and adding, we have

$$X^2 + Y^2 + Z^2 = R^2 \qquad (52)$$

If $R = 0$ equations (51) reduce to (50).

When the forces are in equilibrium any one of the F-*forces* may be considered as a resultant (reversed) of all the others. Equations (50) are therefore general for concurring forces.

The values of the angles α, β, γ, etc., may be determined by drawing a line *from the origin* parallel to and in *the direction of the action of the force*, and measuring the angles from the axes to the line as in Analytical Geometry. The forces may always be considered as positive, and hence the signs of the terms in (50) will be the same as those of the trigonometrical functions. In Fig. 23 the line Oa is parallel to F_2, and the corresponding angles which it makes with the axes are indicated.

If all the forces are in the plane $x\,y$ then γ_1, γ_2, etc. $= 90°$, and (50) becomes

$$\left. \begin{array}{l} X = \Sigma F \cos \alpha = 0; \\ Y = \Sigma F \cos \beta = 0. \end{array} \right\} \qquad (53)$$

CONSTRAINED EQUILIBRIUM.

48. A body is constrained when it is prevented from moving freely under the action of applied forces.

If a particle is constrained to remain at rest on a surface under the action of any number of concurring forces, the resultant of all the applied forces must be in the direction of the normal to the surface at that point.

For, if the resultant were inclined to the normal, it could be resolved into two components, one of which would be

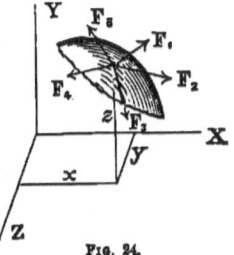

Fig. 24.

tangential, and would produce motion; and the other normal, which would be resisted by the surface.

Let $N =$ the normal reaction of the surface, which will be equal and opposite to the resultant of all the impressed forces;
$\theta_x =$ the angle (N, x);
$\theta_y =$ the angle (N, y);
$\theta_z =$ the angle (N, z);
$L = \phi(x, y, z) = 0$, be the functional equation of the surface; and
F_1, F_2, F_3, etc., be the impressed forces.

Then from (51) and (52), we have

$$\left. \begin{array}{l} X = N \cos \theta_x; \\ Y = N \cos \theta_y; \\ Z = N \cos \theta_z; \\ X^2 + Y^2 + Z^2 = N^2. \end{array} \right\} \quad (54)$$

From Calculus we have

$$\cos \theta_x = \frac{1}{\sqrt{1 + \left(\frac{dx}{dy}\right)^2 + \left(\frac{dx}{dz}\right)^2}}$$

$$= \frac{\left(\frac{dL}{dx}\right)}{\sqrt{\left(\frac{dL}{dx}\right)^2 + \left(\frac{dL}{dy}\right)^2 + \left(\frac{dL}{dz}\right)^2}} \quad (55).$$

and similarly for $\cos \theta_y$ and $\cos \theta_z$.

These values in (54) readily give

$$\frac{X}{\left(\frac{dL}{dx}\right)} = \frac{Y}{\left(\frac{dL}{dy}\right)} = \frac{Z}{\left(\frac{dL}{dz}\right)}. \quad (56)$$

After substituting the values of $\cos \theta_x$, $\cos \theta_y$, and $\cos \theta_z$ in

(54), multiply the first equation by dx, the second by dy, the third by dz, add the results, and reduce by the equation

$$\left(\frac{dL}{dx}\right) dx + \left(\frac{dL}{dy}\right) dy + \left(\frac{dL}{dz}\right) dz = 0;$$

which is the total differential of the equation $L = 0$; and we have

$$Xdx + Ydy + Zdz = 0. \qquad (57)$$

Equations (56) give two independent simultaneous equations which, combined with the equation of the surface, will determine the point of equilibrium if there be one. Equation (57) is one of condition which will be satisfied if there be equilibrium.

To deduce (55) let $f(x', y') = 0$, and $f'(x', z') = 0$, be the equations of the normal to the surface at the point where the forces are applied. In Fig. 25 let Oa be drawn through the origin of co-ordinates parallel to the required normal, then will dx', dy' and dz' be directly proportional to the co-ordinates of a;

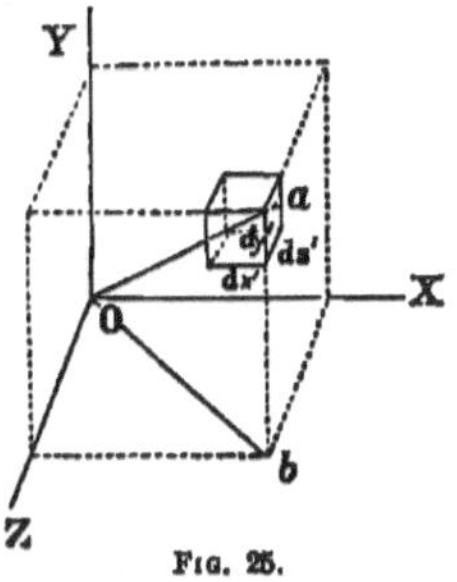

$$\therefore \cos aOx = \cos \theta_x = \frac{x}{Oa}$$

$$= \frac{dx'}{\sqrt{dx'^2 + dy'^2 + dz'^2}}$$

$$= \frac{1}{\sqrt{1 + \left(\frac{dy'}{dx'}\right)^2 + \left(\frac{dz'}{dx'}\right)^2}}.$$

Fig. 25.

But the normal is perpendicular to the tangent plane, and hence the projections of the normal are perpendicular to the traces of the tangent plane. The Equation of Condition of Perpendicularity is of the form $1 + aa' = 0$ (See Analytical Geometry); in which $a' = \frac{dy'}{dx'}$, and $a = \frac{dy}{dx}$; the latter of which is deduced from the equation of the surface;

$$\therefore 1 + \frac{dy'}{dx'} \cdot \frac{dy}{dx} = 0; \text{ and similarly}$$

$$1 + \frac{dz'}{dx'} \cdot \frac{dz}{dx} = 0;$$

hence

$$\frac{dy'}{dx'} = -\frac{dx}{dy} = \frac{\left(\frac{dL}{dy}\right)}{\left(\frac{dL}{dx}\right)}; \text{ and } \frac{dz'}{dx'} = -\frac{dx}{dz} = \frac{\left(\frac{dL}{dz}\right)}{\left(\frac{dL}{dx}\right)};$$

the last terms of which contain the partial differential co-efficients deduced from the equation of the surface. These, substituted in the value of $\cos \theta_x$ above, and reduced, give equation (55).

CONSTRAINED EQUILIBRIUM IN A PLANE.

49. If all the forces are in the plane of a curve, let the plane yx coincide with that plane; then $Z = 0$ and (56) becomes

$$\left. \begin{array}{c} \dfrac{X}{\left(\dfrac{dL}{dx}\right)} = \dfrac{Y}{\left(\dfrac{dL}{dy}\right)}; \\ \text{or, } Xdx = -Ydy; \end{array} \right\} \quad (58)$$

$$\text{and, } Xdx + Ydy = 0; \quad (59)$$

in which the first of (58) may be used when the equation of the curve is given as an implicit function; and the second of (58), or (59), when the equation is an explicit function.

When the particle is not constrained it has three degrees of freedom (equations (50)); when confined to a surface, two degrees (equations (56)); and when confined to a plane curve, only one degree (equation (58)).

EXAMPLES.

1. A body is suspended vertically by a cord which passes over a pulley and is attached to another weight which rests upon a plane; required the position of equilibrium.

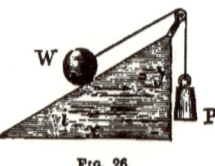

Fig. 26.

In Fig. 26, let the pulley be at the upper end of the plane and the cord and plane perfectly smooth. The weight P is equivalent to a force which acts parallel to the plane, tending to move the weight W up it.

Let W = the weight on the plane, which acts vertically downwards;

P = the weight suspended by the cord;

i = the inclination of the plane to the horizontal; and

$L = -y + ax + b = 0$, be the equation of the plane.

Then
$$X = P \cos i;$$
$$Y = -W + P \sin i;$$

$$a = \tan i = \frac{\sin i}{\cos i};$$

$$\left(\frac{dL}{dy}\right) = -1, \text{ and } \left(\frac{dL}{dx}\right) = a;$$

and these in (58) give

$$P = W \sin i;$$

which only establishes a relation between the constants, and thus determines the relation which must exist in order that there may be equilibrium; and since the variable co-ordinates do not appear, there will be equilibrium at all points along the plane when $P = W \sin i$.

The equation of the line, given explicitly, is

$$y = ax + b;$$
$$\therefore dy = a\, dx;$$

which in the 2nd of (58), or in (59), gives, $P = W \sin i$ as before.

2. Two weights P and W are fastened to the ends of a cord, which passes over a pulley O; the weight W rests upon a vertical plane curve, and P hangs freely; required the position of equilibrium.

The applied forces at W are the weight W, acting vertically downward; the tension P on the string; and the normal reaction of the curve.

[Consider the weight W and pulley O as reduced to points.]

Let $r = OW$; $\theta = WOA$; $y = OA$;

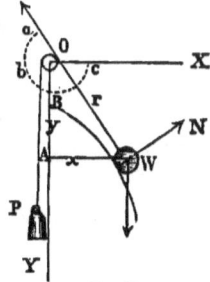

Fig. 27.

Then,
$$\sin\theta = \frac{x}{r}; \quad \cos\theta = \frac{y}{r}; \quad r^2 = x^2 + y^2;$$

$$Y = W + P\cos aOB = W - P\cos\theta;$$
$$X = P\cos cba = -P\sin\theta;$$

and (59) becomes
$$-P\sin\theta\, dx + (W - P\cos\theta)\, dy = 0;$$
or,
$$Wdy = P\frac{xdx + ydy}{r} = Pdr;$$

which integrated gives
$$Wy = Pr + C; \qquad (a)$$

and this, combined with the equation of the curve, gives the required co-ordinates.

3. Let the given curve be a parabola, in which the origin is at the focus.

The equation of the curve will be $x^2 = 2p(y + \frac{1}{2}p)$; but it is unnecessary to use it, since, by a well-known property of the curve, *r equals the distance from the point on the curve to the directrix* $= y + p$. Substituting this value of r in equation (a), we have
$$Wy = P(y + p) + C.$$

To find C we observe that when the weight W, Fig. 27, is in the horizontal line through the focus, $y = 0 \therefore C = -Pp$; and this value substituted above gives
$$Wy = Py; \text{ or } W = P;$$

from which it appears that if $W = P$, the weights will be in equilibrium at all points on the curve. The result holds true when the parabola reduces to the particular case of two vertical straight lines passing through the focus.

This problem may also be proved by observing that the normal bisects the angle formed by the radius vector and diameter passing through any point, and, hence, the forces along the diameter and radius vector, which are the components, must equal each other.

4. Let the curve be a circle in which the distance of O from the centre is a; and the equation of the circle is

$$(a-y)^2 + x^2 = R^2.$$

$$Ans.\ r = \frac{P}{W}a.$$

5. Let the curve be an hyperbola, having the origin of coördinates at the centre of the hyperbola.

The equation of the curve will be $a^2x^2 - b^2y^2 = -a^2b^2$, and if e is the eccentricity, we have

$$y = \frac{bW}{e(W^2 - e^2P^2)^{\frac{1}{2}}}.$$

6. Required the curve such that the weight W may be in equilibrium with any weight P at all points of the curve.

This requires that the relation between y and r (or y and x) in equation (d) shall be true for all assumed values of W and P.

In Fig. 27, let $OB = a$, be the distance of the origin of coördinates from the vertex of the curve, then when W is at B, we have $y = r = a$, which, in equation (a), gives

$$C = (W - P)a;$$

and from the figure we have

$$y = r \cos \theta;$$

which values in (a) finally give

$$r = \frac{1 - \dfrac{W}{P}}{1 - \dfrac{W}{P}\cos\theta}\, a,$$

which is the equation of a conic of which the focus is at the pole O.

(Discuss the equation and determine all the species of the conic.)

7. A particle is placed on the concave surface of a smooth sphere and acted upon by gravity, and also by a repulsive

force, which varies inversely as the square of the distance from the lowest point of the sphere; find the position of equilibrium of the particle.

Take the lowest point of the sphere for the origin of coördinates, y positive upwards, and the equation of the surface will be

$$L = x^2 + y^2 + z^2 - 2Ry = 0.$$

Let r be the distance of the particle from the lowest point; then

$$r^2 = x^2 + y^2 + z^2 = 2Ry. \qquad (b)$$

Let μ be the measure of the repulsive force at a unit's distance; then the forces will be

$$\frac{\mu}{r^2} = \frac{\mu}{2Ry}, \text{ and } mg = w = \text{the weight of the particle.}$$

$$\therefore X = \frac{\mu}{2Ry} \cdot \frac{x}{r}, \quad Y = \frac{\mu}{2Ry} \cdot \frac{y}{r} - w, \quad Z = \frac{\mu}{2Ry} \cdot \frac{z}{r};$$

which in (56) give, after reduction,

$$y = \frac{\mu^{\frac{1}{3}}}{2R^{\frac{1}{3}} w^{\frac{1}{3}}};$$

which in (b) gives, $r^3 = \dfrac{\mu}{w} R$.

To see if these values satisfy equation (57), substitute in it the values of X, Y, Z, and the final values of y and r, and we find,

$$x\,dx + y\,dy - R\,dy + z\,dz = 0;$$

which is the differential of equation (b), and hence is true.

[This is the theory of *the Electroscope.*]

8. A particle on the surface of an ellipsoid is attracted by forces which vary directly as its distance from the principa planes of section; determine the position of equilibrium.

Let
$$L = \phi(x, y, z) = \frac{x^2}{a^2} + \frac{y^2}{b^2} + \frac{z^2}{c^2} - 1 = 0,$$

be the equation of the surface;

$$\therefore \left(\frac{dL}{dx}\right) = \frac{2xdx}{a^2}, \quad \left(\frac{dL}{dy}\right) = \frac{2ydy}{b^2}, \quad \left(\frac{dL}{dz}\right) = \frac{2zdz}{c^2};$$

and let the x, y, and z-*components* of the forces be respectively,

$$X = -\mu_1 x, \quad Y = -\mu_2 y, \quad Z = -\mu_3 z;$$

and (56) will give,

$$\mu_1 a^2 = \mu_2 b^2 = \mu_3 c^2;$$

which simply establishes a relation between the constants; and hence when this relation exists the particle may be at rest at any point on the surface.

The result may be put in the form,

$$\frac{\mu_1}{a^{-2}} = \frac{\mu_2}{b^{-2}} = \frac{\mu_3}{c^{-2}} = \frac{\mu_1 + \mu_2 + \mu_3}{a^{-2} + b^{-2} + c^{-2}}$$

MOMENTS OF FORCES.

50. Def. *The moment of a force in reference to a point is the product arising from multiplying the force by the perpendicular distance of the action-line of the force from the point.*

Thus, in Fig. 28, if O is the point from which the perpendicular is drawn, F the force, and Oa the perpendicular, then the moment of F is

$$F.Oa = Ff;$$

in which f is the perpendicular Oa.

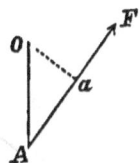

Fig. 28.

51. Nature of a moment. The moment of a force measures the turning or twisting effect of a force. Thus, in Fig. 28, if the particle upon which the force F acts is at A, and if we conceive that the point O is rigidly connected to A, the force will tend to move the particle about O, and it is evident that this effect varies directly as F. If the action-line of F passed through O it would have no tendency to move the particle about that point, and the greater its distance from that point the greater will be its effect, and it will vary directly as that distance; hence, *the measure of the effect of a moment varies as the product of the force and perpendicular;* or as

$$cFf;$$

where c is a constant. But as c may be chosen arbitrarily, we make it equal to unity, and have simply Ff, as given above.

52. DEF. The point O from which the perpendiculars are drawn is chosen arbitrarily, and is called the *origin of moments*. When the system is referred to rectangular coördinates, the *origin of moments* may, or may not, coincide with the origin of coördinates. The solution of many problems is simplified by taking the origin of moments at a particular point.

53. THE LEVER ARM, or, simply, *the arm*, of a force is the perpendicular from the origin of moments to the action-line of the force. Thus, in Fig. 29, Oa is the arm of the force F_1; Oc that of the force F_3, etc. *Generally, the arm is the perpendicular distance of the action-line from the axis* about which the system is supposed to turn.

54. THE SIGN OF A MOMENT is considered *positive* if it tends to turn the system in a direction opposite to that of the hands of a watch; and *negative*, if in the opposite direction. This is arbitrary, and the opposite directions may be chosen with equal propriety; but this agrees with the direction in which the angle is computed in plane trigonometry. *Generally* we shall consider those moments as *positive* which tend to turn the system in the direction indicated by the natural order of the letters, that is, *positive* from $+x$ to $+y$; from $+y$ to $+z$; then from $+z$ to $+x$; and *negative* in the reverse direction.

The value of a moment may be represented by a straight line drawn from the origin and along the line about which rotation tends to take place, *in one direction for a positive value,* and in the opposite direction for a negative one.

55. THE COMPOSITION AND RESOLUTION of moments may be effected in substantially the same manner as for forces. They may be added, or subtracted, or compounded, so that a resultant moment shall produce the same effect as any number of single moments. The general proof of this proposition is given in the next Chapter.

56. A MOMENT AXIS is a line passing through the origin of moments and perpendicular to the plane of the force and arm.

57. THE MOMENT OF A FORCE REFERRED TO A MOMENT AXIS *is the product of the force into the perpendicular distance of the force from the axis.*

If, in Fig. 29, a line is drawn through O perpendicular to the plane of the force and arm, it will be a moment axis, and the turning effect of F_1 upon that axis will be the same wherever applied, providing that its arm Oa remains constant.

If the force is not perpendicular to the arbitrarily chosen axis, it may be resolved into two forces, one of which will be perpendicular (but need not intersect it) and the other parallel to the axis. The moment of the former component will be the same as that given above, but the latter will have no moment in reference to *that* axis although it may have a moment in reference to another axis perpendicular to the former.

58. THE MOMENT OF A FORCE IN REFERENCE TO A PLANE TO WHICH IT IS PARALLEL *is the product of the force into the distance of its action-line from the plane.*

59. *If any number of concurring forces are in equilibrium the algebraic sum of their moments will be zero.*

Let F_1, F_2, F_3, etc., Fig. 29, be the forces acting upon a particle at A; and O the assumed origin of moments. Join O and A, and let fall the perpendiculars Oa, Ob, Oc, etc., upon the action-lines of the respective forces, and let

$Oa = f_1$; $Ob = f_2$; $Oc = f_3$; etc.

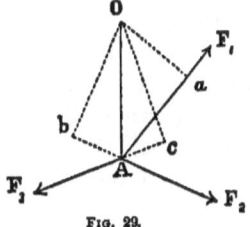

FIG. 29.

Resolve the forces perpendicularly to the line OA; and since they are in equilibrium, the algebraic sum of these components will be zero; hence,

$$F_1 \sin OAF_1 + F_2 \sin OAF_2 + F_3 \sin OAF_3 + \text{etc.} = 0;$$

$$\text{or, } F_1 \frac{Oa}{OA} + F_2 \frac{Ob}{OA} + F_3 \frac{Oc}{OA} + \text{etc.} = 0.$$

Multiply by OA, and we have

$$F_1 Oa + F_2 Ob + F_3 Oc + \text{etc.} = 0;$$

$$\text{or, } F_1 f_1 + F_2 f_2 + F_3 f_3 + \text{etc.} = \Sigma F f = 0. \quad (60)$$

It is evident that any one of these moments may be taken as the resultant of all the others.

MOMENTS OF CONCURRING FORCES WHEN THE SYSTEM IS REFERRED TO RECTANGULAR AXES.

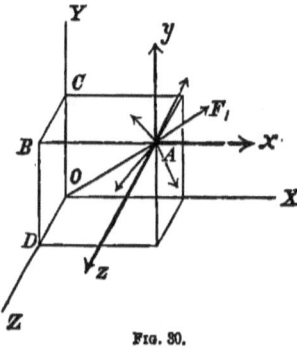

Fig. 30.

60. Let A, Fig. 30, be the point of application of the forces F_1, F_2, F_3, etc., and O the origin of coördinates, and also the origin of moments. Let x, y, and z be the coördinates of the point A. Resolving the forces parallel to the coördinate axes, we have, from equation (50),

$$X = \Sigma F \cos \alpha;$$
$$Y = \Sigma F \cos \beta;$$
$$Z = \Sigma F \cos \gamma.$$

The X-forces prolonged will meet the *plane* of yz in B; and will tend to turn the system about the axis of y, in reference to which it has the arm $BC = z$; and also about z, in reference to which it has the arm $BD = y$. Hence, employing the notation already established, we have for the moment of the sum of the components parallel to x,

$$- Xy, \text{ and } + Xz.$$

Similarly for the *y-components* we find the moments,

$$+ Yx, \text{ and} - Yz;$$

and for the *z-components*,

$$- Zx, \text{ and} + Zy.$$

The moment Xy tends to turn the system one way about the axis of z, and Yx tends to turn it about the same axis, but in the opposite direction; and hence, the combined effect of the two will be their algebraic sum; or

$$Yx - Xy.$$

But since there is equilibrium the sum will be zero. Combining the others in the same manner, we have, for *the moments of concurring forces*, in equilibrium:

$$\left. \begin{array}{l} \text{In reference to the axis of } x \ldots Zy - Yz = 0; \\ \text{} y \ldots Xz - Zx = 0; \\ \text{} z \ldots Yx - Xy = 0. \end{array} \right\} \quad (61)$$

The third equation may be found by eliminating z from the other two; hence, when X, Y, and Z are known, they are the equations of a straight line; and are *the equations of the resultant*.

If the origin of moments be at some other point, whose coördinates are x', y', and z'; and the coördinates of the point A in reference to the origin of moments be x_1, y_1, and z_1; then will the lever arms be

$$x_1 = x - x'; \ y_1 = y - y'; \text{ and } z_1 = z - z'.$$

When the system is referred to rectangular coördinates the arm of the force, referred to the z-axis, is

$$y \cos a - x \cos \beta,$$

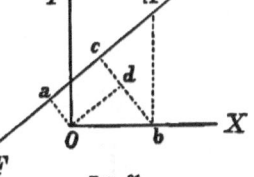

Fig. 31.

in which y and x are the coördinates of *any point* of the action-line of the force; and a is the angle which the action-line makes with the axis of x, and β the angle which it makes with y.

In Fig. 31, let AF be the action-line of the force F, O the origin of coördinates, A any point in the line AF, of which the coördinates $x = Ob$, and $y = Ab$. Draw Oa and bc perpendicular to AF, and Od from O parallel to AF. The origin of moments being at O, Oa will be the arm of the force.

We have
$$dOb = a = cbA,$$
$$cAb = \beta = Obd,$$
$$cb = y \cos a,$$
$$db = x \cos \beta;$$
$$\therefore aO = cb - db = y \cos a - x \cos \beta. \qquad (61a)$$

If there are three coördinate axes, this will be the arm in reference to the axis of z; and if there be many forces, the sum of their moments in reference to that axis, will be

$$\Sigma F (y \cos a - x \cos \beta).$$

Examples.

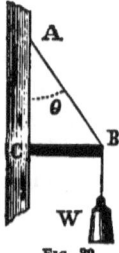
Fig. 32.

1. A weight W is attached to a string, which is secured at A, Fig. 32, and is pushed from a vertical by a strut CB; required the pressure F on BC when the angle CAB is θ.

The forces which concur at B are the weight W, the pressure F, and the tension of the string AB. Take the origin of moments at A, and we have

$$- W.BC + F.AC + tension \times 0 = 0;$$

$$\therefore F = W\frac{BC}{AC} = W \tan \theta.$$

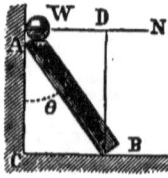
Fig. 33.

2. A brace, AB, rests against a vertical wall and upon a horizontal plane, and supports a weight W at its upper end; required the compression upon the brace and the thrust at A when the angle CAB is θ; the end B being held by a string BC.

The concurring forces at A, are W, acting vertically downward, the reaction of the wall N acting horizontally, and the reaction of the brace F.

Take the origin of moments at B, we have

$$- N.DB + W.CB + F.0 = 0;$$

$$\therefore N = W \tan \theta.$$

Taking the origin of moments at D, we have

$$W. AD - F. DB \sin \theta + N.0 = 0;$$

$$\therefore F = W \sec \theta.$$

3. A rod whose length is $BC = l$ is secured at a point B, in a horizontal plane, and the end C is held up by a cord AC so that the angle BAC is θ, and the distance $AB = a$; required the tension on AC and compression on BC, due to a weight W applied at C.

$$Ans. \quad Compression = \frac{l}{a} W \cot \theta.$$

4. A cord whose length $AC = l$ is secured at two points in a horizontal line, and a weight W is suspended from it at B; required the tension on each part of the cord.

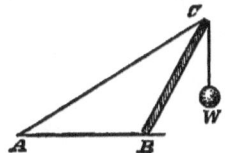 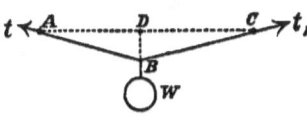

CHAPTER III.

PARALLEL FORCES.

61. Bodies are extended masses, and forces may be applied at any or all of their points, and act in all conceivable directions, as in Fig. 34.

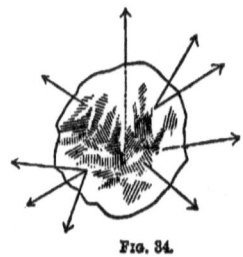

Fig. 34.

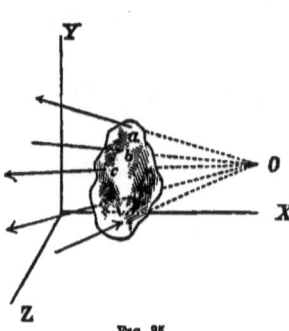

Fig. 35.

62. Suppose that the action-lines of all the forces are parallel to each other. This is a special case of concurrent forces, in which the point of meeting of the action-lines is at an infinite distance. In Fig. 35, let the points a, b, c, etc., which are on the action-lines of the forces and within the body, be the points of application of the forces, and O the point where they would meet if prolonged. If the point O recedes from the body, while the points of application a, b, c, etc. remain fixed, the action-lines of the forces will approach parallelism with each other, and at the limit will be parallel.

63. Resultant of parallel forces. The forces being parallel, the angles which they make with the respective axes, including those of the resultant, will be equal to each other Hence,

$$a = a_1 = a_2 = a_3, \text{ etc.} = a;$$
$$b = \beta_1 = \beta_2 = \beta_3, \text{ etc.} = \beta;$$
$$c = \gamma_1 = \gamma_2 = \gamma_3, \text{ etc.} = \gamma;$$

and these, in equations (50) and (51), give

$$X = R \cos a = (F_1 + F_2 + F_3 + \text{etc.}) \cos a.$$
$$Y = R \cos \beta = (F_1 + F_2 + F_3 + \text{etc.}) \cos \beta. \quad (62)$$
$$Z = R \cos \gamma = (F_1 + F_2 + F_3 + \text{etc.}) \cos \gamma.$$

From either of these, we have

$$R = F_1 + F_2 + F_3 + \text{etc.} = \Sigma F. \quad (63)$$

Hence, *the resultant of parallel forces equals the algebraic sum of the forces.*

From (62), we have

$$R = \sqrt{X^2 + Y^2 + Z^2},$$

which is the same as (52).

MOMENTS OF PARALLEL FORCES.

64. Let F_1, F_2, F_3, etc., be the forces, and x_1, y_1, z_1; x_2, y_2, z_2, etc., be the coördinates of the points of application of the forces respectively (which, as before stated, may be at any point on their action-lines). Then the moments of F_1 will be, according to Article (60),

in reference to the axis of x, $F_1 \cos \gamma \cdot y_1 - F_1 \cos \beta \cdot z_1;$

" " " " " " y, $F_1 \cos a \cdot z_1 - F_1 \cos \gamma \cdot x_1;$

" " " " " " z, $F_1 \cos \beta \cdot x_1 - F_1 \cos a \cdot y_1;$

and similarly for all the other forces. Hence, the *sum of the moments* in reference to the respective axes for equilibrium is,

$$\left. \begin{array}{l} (F_1 y_1 + F_2 y_2 + F_3 y_3 + \text{etc.}) \cos \gamma \\ -(F_1 z_1 + F_2 z_2 + F_3 z_3 + \text{etc.}) \cos \beta \end{array} \right\} = 0;$$

$$\left. \begin{array}{l} (F_1 z_1 + F_2 z_2 + F_3 z_3 + \text{etc.}) \cos a \\ -(F_1 x_1 + F_2 x_2 + F_3 x_3 + \text{etc.}) \cos \gamma \end{array} \right\} = 0;$$

$$\left. \begin{array}{l} (F_1 x_1 + F_2 x_2 + F_3 x_2 + \text{etc.}) \cos \beta \\ -(F_1 y_1 + F_2 y_2 + F_3 y_3 + \text{etc.}) \cos a \end{array} \right\} = 0.$$

These equations will be true for all values of α, β, and γ, if the coefficients of $\cos \alpha$, $\cos \beta$, $\cos \gamma$, are respectively equal to zero; for which case we have

$$\left. \begin{array}{l} F_1 x_1 + F_2 x_2 + F_3 x_3 + \text{etc.} = \Sigma Fx = 0\,; \\ F_1 y_1 + F_2 y_2 + F_3 y_3 + \text{etc.} = \Sigma Fy = 0\,; \\ F_1 z_1 + F_2 z_2 + F_3 z_3 + \text{etc.} = \Sigma Fz = 0\,; \end{array} \right\} \quad (64)$$

from which the coördinates of the point of application of any one of the forces, as F_1, for instance, may be found so as to satisfy these equations, when all the other quantities are given.

Let the given forces have a resultant. Let a force, as F_1, equation (64), equal and opposite to the resultant, be introduced into the system, then will there be equilibrium. Let ΣFx, ΣFy, ΣFz, include the sum of the respective products for all the forces *except* that of the resultant; R be the resultant, and $\bar{x}$, $\bar{y}$, $\bar{z}$, the coördinates of the point of application of the resultant; so chosen as to satisfy equations (64), then we have

$$R\bar{x} - \Sigma Fx = 0\,; \quad R\bar{y} - \Sigma Fy = 0\,; \quad R\bar{z} - \Sigma Fz = 0. \quad (65)$$

Substitute the value of $R = \Sigma F$, in these equations, and we find

$$\bar{x} = \frac{\Sigma Fx}{\Sigma F}\,; \qquad \bar{y} = \frac{\Sigma Fy}{\Sigma F}\,; \qquad \bar{z} = \frac{\Sigma Fz}{\Sigma F}\,; \quad (66)$$

by which the point of application of the resultant becomes known, and, being independent of α, β, and γ, *is a point through which the resultant constantly passes, as the forces are turned about their points of application, the forces constantly retaining their parallelism.* This point is called *the centre of parallel forces.*

65. IF THE SYSTEM CONSISTS OF THREE FORCES ONLY, and are in the plane xy, we have

$$\left. \begin{array}{c} R = F_1 + F_2\,; \\ \bar{x} = \dfrac{F_1 x_1 + F_2 x_2}{F_1 + F_2}\,; \text{ and } \bar{y} = \dfrac{F_1 y_1 + F_2 y_2}{F_1 + F_2}. \end{array} \right\} \quad (67)$$

[66.]

STATICAL COUPLES.

1st. *Consider F_1 and F_2 as positive.*
The resultant will equal the arithmetical sum of the forces. Take the origin at a Fig. 36, where the resultant cuts the axis of x; then $\bar{x} = 0$, and the second of (67) gives

$$F_1 x_1 = - F_2 x_2;$$

and hence, if $F_1 > F_2$, x_2 will exceed x_1; that is, the resultant is nearer the greater force.

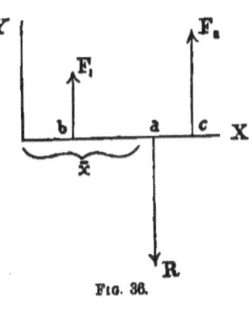

Fig. 36.

2d. *Consider F_2 as negative.*

In this case the resultant equals the difference of the forces. Take the origin at a, Fig. 37, and we have

$$F_1 x_1 = F_2 x_2;$$

and hence both forces are either at the right or left of the resultant.

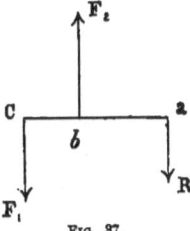

Fig. 37.

3d. Let $F_1 = F_2 = F$, and one of the forces be negative, then

$$R = F - F = 0; \quad \bar{x} = \frac{F(x_1 \pm x_2)}{F - F} = \infty; \text{ and } \bar{y} = \infty; \quad (68)$$

that is, the resultant is zero, while the forces may have a finite moment equal to $F(x_1 \pm x_2)$. Such systems are called

COUPLES.

66. *A couple consists of two equal parallel forces acting in opposite directions at a finite distance from each other.*

A statical couple cannot be equilibrated by a single force. It does not produce translation, but simply rotation. *A couple can be equilibriated only by an equivalent couple.*

Equivalent couples are such as have equal moments.

The resultant of several couples is a single couple which will produce the same effect as the component couples.

67. *The arm of a couple is the perpendicular distance between the action-lines of the forces.* Thus, in Fig. 38, let O be the origin of coördinates, and the axis of x perpendicular to the action-line of F; then will the moment of one force be Fx_1, and of the other Fx_2, and hence the *resultant moment will be*

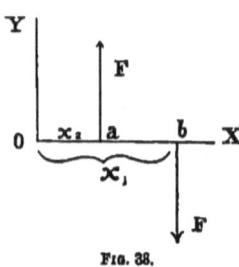

Fig. 38.

$$F(x_1 - x_2) = F.ab; \quad (69)$$

hence, ab is the arm. If the origin of coördinates were between the forces the moments would be $F(x_1 + x_2) = F.ab$ as before. If the origin be at a we have $F.0 + F.ab = F.ab$ as before.

68. THE AXIS OF A STATICAL COUPLE *is any line perpendicular to the plane of the couple.* The length of the axis may be made proportional to the moment of the couple, and placed on one side of the plane when the moment is positive, and on the opposite side when it is negative; and thus completely represent the couple in magnitude and direction.

If couples are in parallel planes, their axes may be so taken that they will conspire, and hence the resultant couple equals the algebraic sum of all the couples.

If the planes of the couples intersect, their axes may intersect.

Let $O = F.ab =$ the moment of one couple;
$O_1 = F_1.a_1b_1 =$ the moment of another couple;
$\theta =$ the angle between their axes; and
$O_R =$ the resultant of the two couples;
then

$$O_R = \sqrt{O^2 + O_1^2 + 2OO_1 \cos\theta};$$

and this resultant may be combined with another and so on until the final resultant is obtained.

EXAMPLES.

1. Three forces represented in magnitude, direction and *position,* by the sides of a triangle, taken in their order, produce a couple.

2. If three forces are represented in magnitude and position by the sides of a triangle, but whose directions do not follow the *order* of the sides; show that they will have a single resultant.

3. On a straight rod are suspended several weights; $F_1 = 5$ lbs., $F_2 = 15$ lbs., $F_3 = 7$ lbs., $F_4 = 6$ lbs., $F_5 = 9$ lbs., at distances $AB = 3$ ft., $BD = 6$ ft., $DE = 5$ ft., and $EF = 4$ ft.; required the distance AC at which a fulcrum must be placed so that the weights will balance on it; also required the pressure upon it.

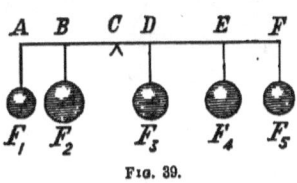

Fig. 39.

4. The whole length of the beam of a false balance is 2 feet 6 inches. A body placed in one scale balances 6 lbs. in the other, but when placed in the other scale it balances 8 lbs.; required the true weight of the body, and the lengths of the arms of the balance.

5. A triangle in the horizontal plane x, y has weights at the several angles which are proportional respectively to the opposite sides of the triangle; required the coördinates of the centre of the forces.

Let x_1, y_1 be the coördinates of A,
x_2, y_2 of B; x_3, y_3 of C;
$\bar{x}$, $\bar{y}$ of the point of application of the resultant;
then we have
$$(a + b + c)\bar{x} = ax_1 + bx_2 + cx_3; \text{ and}$$
$$(a + b + c)\bar{y} = ay_1 + by_2 + cy_3.$$

6. If weights in the proportion of 1, 2, 3, 4, 5, 6, 7 and 8 are suspended from the respective angles of a parallelopiped; required the point of application of the resultant.

7. Several couples in a plane, whose forces are parallel, are applied to a rigid right line, as in Fig. 40; required the resultant couple.

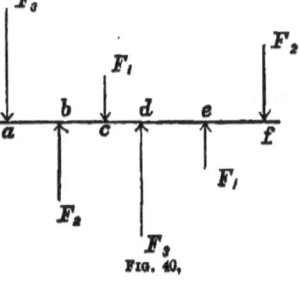

Fig. 40.

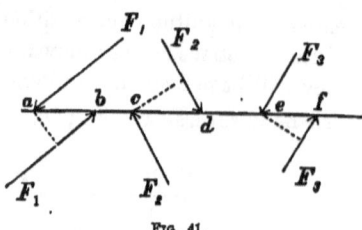

Fig. 41.

8. Several couples in a plane, whose respective arms are not parallel, as in Fig. 41, act upon a rigid right line; required the resultant couple.

CENTRE OF GRAVITY OF BODIES.

69. The action-lines of the force of gravity are normal to the surface of the earth, but, for those bodies which we shall here consider, their convergence will be so small, that we may consider them as parallel. We may also consider the force as the same at all points of the body.

The *centre of gravity of a body* is the point of application of the resultant of the force of gravity as it acts upon every particle of the body. It is the centre of parallel forces. If this point be supported the body will be supported, and if the body be turned about this point it will remain constantly in the centre of the parallel forces.

Let $M =$ the mass of a body;
$m =$ the mass of an infinitesimal element;
$V =$ the volume of the body;
$D =$ the density at the point whose coördinates are x, y, and z;
$R = W =$ the resultant of gravity, which is the weight; and
$\bar{x}$, $\bar{y}$, and $\bar{z}$ be the coördinates of the centre of gravity.

We have, according to equations (63) and (20),
$$R = W = \Sigma gm = M \times g;$$
and (65) becomes
$$\left. \begin{array}{l} \bar{x} \Sigma gm = \Sigma gmx; \text{ or } M\bar{x} = \Sigma mx; \\ \bar{y} \Sigma gm = \Sigma gmy; \text{ or } M\bar{y} = \Sigma my; \\ \bar{z} \Sigma gm = \Sigma gmz; \text{ or } M\bar{z} = \Sigma mz. \end{array} \right\} \quad (70)$$

If the density is a continuous function of the coördinates of the body we may integrate the preceding expressions. The complete solution will sometimes require two or three integrations, depending upon the character of the problem; but, using only one integral sign, (22) and (70) become

$$\left. \begin{array}{l} M = \int D \, dV; \\ \bar{x} \int D \, dV = \int Dx \, dV; \\ \bar{y} \int D \, dV = \int Dy \, dV; \\ \bar{z} \int D \, dV = \int Dz \, dV. \end{array} \right\} \quad (71)$$

If the origin of coördinates be at the centre of gravity, then

$$\bar{x} = 0; \ \bar{y} = 0; \ \bar{z} = 0;$$

and hence,

$$\Sigma mx = \int Dx \, dV = 0; \quad (71a)$$

and similarly for the other values.

If D be constant, this becomes

$$\int_{\prime\prime}^{\prime\prime} x \, dV = 0; \quad (71b)$$

the limits of integration including the whole body.

If the mass is homogeneous, the density is uniform, and D being cancelled in the preceding equations, we have

$$\left. \begin{array}{l} \bar{x} = \dfrac{\int x \, dV}{V}; \\ \\ \bar{y} = \dfrac{\int y \, dV}{V}; \\ \\ \bar{z} = \dfrac{\int z \, dV}{V}. \end{array} \right\} \quad (72)$$

Many solutions may be simplified by observing the following principles:

1. *If the body has an axis of symmetry the centre of gravity will be on that axis.*

2. *If the body has a plane of symmetry the centre of gravity will be in that plane.*

3. *If the body has two or more axes of symmetry the centre of gravity will be at their intersection.*

Hence, the centre of gravity of a physical straight line of uniform density will be at the middle of its length; that of the circumference of a circle at the centre of the circle; that of the circumference of an ellipse at the centre of the ellipse; of the area of a circle, of the area of an ellipse, of a regular polygon, at the geometrical centre of the figures. Similarly the centre of gravity of a triangle will be in the line joining the vertex with the centre of gravity of the base; of a pyramid or cone in the line joining the apex with the centre of gravity of the base.

There is a certain inconsistency in speaking of the centre of gravity of geometrical lines, surfaces, and volumes; and when they are used, it should be understood that a *line* is a *physical* or *material line* whose section may be infinitesimal; a *surface* is a *material section*, or thin plate, or thin shell; and a *volume* is a mass, however attenuated it may be.

When a body has an axis of symmetry, the axis of x may be made to coincide with it, and only the first of the preceding equations will be necessary. If it has a plane of symmetry, the plane $x\,y$ may be made to coincide with it, and only the first and second will be necessary.

70.   *Centre of gravity of material lines.*

Let $k=$ the transverse section of the line, and
$ds=$ an element of the length,
then
$$dV = kds;$$
and (71) becomes

$$\left. \begin{array}{l} \bar{x}\!\int\!Dkds = \int\!Dkxds\,; \\ \bar{y}\!\int\!Dkds = \int\!Dkyds\,; \\ \bar{z}\!\int\!Dkds = \int\!Dkzds. \end{array} \right\} \quad (73)$$

[70.] OF LINES. 95

If the transverse section and the density are uniform, we have

$$\left.\begin{array}{c}\bar{x} = \dfrac{\int x ds}{s};\\[4pt] \bar{y} = \dfrac{\int y ds}{s};\\[4pt] \bar{z} = \dfrac{\int z ds}{s}.\end{array}\right\} \quad (74)$$

The centre of gravity will sometimes be outside of the line or body, and hence, if it is to be supported at that point, we must conceive it to be rigidly connected with the body by lines which are without weight.

EXAMPLES.

1. Find the centre of gravity of a straight fine wire of uniform section in which the density varies directly as the distance from one end.

Let the axis of x coincide with the line, and the origin be taken at the end where the density is zero. Let δ be the density at the point where $x = 1$; then for any other point it will be $D = \delta x$; and substituting in the first of (73), also making $ds = dx$, we have

$$\bar{x}\int_0^a \delta x dx = \int_0^a \delta x^2 dx; \quad \therefore \bar{x} = \tfrac{2}{3}a.$$

This corresponds with the distance of the centre of gravity of a triangle from the vertex.

Fig. 42.   Fig. 43.

2. Find the centre of gravity of a cone or pyramid, whether

right or oblique, and whether the base be regular or irregular.

Draw a line from the apex to the centre of gravity of the base, and conceive that all sections parallel to the base are reduced to this central line. The problem is then reduced to finding the centre of gravity of a physical line in which the density increases as the square of the distance from one end.

$$Ans.\ \bar{x} = \tfrac{3}{4}a.$$

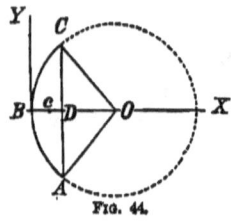

Fig. 44.

3. To find the centre of gravity of a circular arc.

Let the axis of x pass through the centre of the arc, B, and the centre of the circle O; then $\overline{Y} = 0$. Take the origin at B;

and let
$$x = BD,$$
$$y = DC,$$
$$2s = \text{the arc } ABC, \text{ and}$$
$$r = OC = \text{the radius of the circle;}$$

then,
$$y^2 = 2rx - x^2.$$

Differentiate, and we have
$$y\,dy = r\,dx - x\,dx;$$
$$\therefore \frac{dy}{r-x} = \frac{dx}{y} \text{ which } = \frac{ds}{r};$$

hence, the first of (74) gives

$$\bar{x} = \frac{r\displaystyle\int_0^y \frac{x\,dx}{\sqrt{2rx - x^2}}}{s} = \frac{r}{s}\left[-\sqrt{2rx - x^2} + s\right]_0^y = r - \frac{ry}{s};$$

which is the distance Bc. Then $cO = r - \left(r - \dfrac{ry}{s}\right) = \dfrac{ry}{s}$,

hence, the distance of the centre of gravity of an arc from the centre of the circle is a fourth proportional to the arc, the radius, and the chord of the arc.

71. *Centre of gravity of plane surfaces.*

Let the coördinate plane xy coincide with the surface; then $dV = dxdy$; $\therefore V = \iint dxdy = \int ydx$ or $\int xdy$; and (71) becomes

$$\bar{x}\iint D\,dxdy = \iint D\,xdxdy ;$$
$$\bar{y}\iint D\,dxdy = \iint D\,ydxdy. \quad (75)$$

The integrals are definite, including the whole area. The order of integration is immaterial, but after the first integration the limits must be determined from the conditions of the problem. If D is constant and the integral is made in respect to y, we have

$$\bar{x} = \frac{\int yxdx}{\int ydx};$$
$$\bar{y} = \frac{\frac{1}{2}\int y^2 dx}{\int ydx}; \quad (76)$$

and if x be an axis of symmetry, the first of these equations will be sufficient.

If the surface is referred to polar coördinates, then $dV = \rho d\rho d\theta$, and $x = \rho\cos\theta$, $y = \rho\sin\theta$, and (71) becomes

$$\bar{x} = \frac{\iint D\rho^2\cos\theta d\rho d\theta}{\iint D\rho d\rho d\theta};$$
$$\bar{y} = \frac{\iint D\rho^2\sin\theta d\rho d\theta}{\iint D\rho d\rho d\theta}. \quad (77)$$

EXAMPLES.

1. Find the centre of gravity of a semi-parabola whose equation is $y^2 = 2px$.

Equations (76) become

$$\bar{x} = \frac{\int_0^x \sqrt{2p}\, x^{\frac{3}{2}} dx}{\frac{2}{3}xy} = \frac{3}{5}x;$$

$$\bar{y} = \frac{\frac{1}{2}\int_0^y 2pxdx}{\frac{2}{3}xy} = \frac{3}{8}y.$$

7

2. Find the centre of gravity of a quadrant of a circle in which the density increases directly as the distance from the centre.

Let $\delta =$ the density at a unit's distance from the centre; then
$$D = \delta\rho \text{ at a distance } \rho;$$
and (77) becomes

$$\bar{x} = \frac{\delta \int_0^{\frac{1}{2}\pi} \int_0^r \rho^3 \cos\theta \, d\rho \, d\theta}{\delta \int_0^{\frac{1}{2}\pi} \int_0^r \rho^2 \, d\rho \, d\theta} = \tfrac{2}{3}\frac{r}{\pi} = \bar{y}.$$

72. *Centre of gravity of curved surfaces.*

We have for an element of the area
$dV = dx\, dy \times$ sec. *of the angle between the tangent plane and the coördinate plane xy;* or,
$$dV = \text{sec. } \theta \times dx\, dy;$$

$$\therefore V = \iint \sqrt{1 + \left(\frac{dz}{dy}\right)^2 + \left(\frac{dz}{dx}\right)^2}\, dx\, dy;$$

or, in terms of partial differential coefficients

$$V = \iint \frac{\sqrt{\left(\frac{dL}{dx}\right)^2 + \left(\frac{dL}{dy}\right)^2 + \left(\frac{dL}{dz}\right)^2}}{\left(\frac{dL}{dz}\right)}\, dx\, dy;$$

and, for a homogeneous surface, (72) becomes

$$\left.\begin{aligned} \bar{x} &= \frac{\iint x\, dV}{V}; \\ \bar{y} &= \frac{\iint y\, dV}{V}; \\ \bar{z} &= \frac{\iint z\, dV}{V}; \end{aligned}\right\} \quad (78)$$

or, the surface may be referred to polar coördinates.

[72.] OF DOUBLE CURVED SURFACES. 99

If the surface is one of revolution, let x coincide with the axis of revolution, then

$$\bar{x} \int \pi y\, ds = \int \pi y x\, ds. \qquad (78a)$$

EXAMPLE.

Find the centre of gravity of one-eighth of the surface of a sphere contained within three principal planes.

Let the equation of the sphere be

$$L = x^2 + y^2 + z^2 - R^2 = 0;$$

then

$$\frac{dL}{dx} = 2x; \quad \frac{dL}{dy} = 2y; \quad \frac{dL}{dz} = 2z;$$

and the first of (78) becomes

$$\bar{x} = \frac{\iint x \frac{\sqrt{4x^2 + 4y^2 + 4z^2}}{2z}\, dx\, dy}{\tfrac{1}{2}\pi R^2}$$

$$= \frac{2}{\pi R} \iint \frac{x}{z}\, dx\, dy$$

$$= \frac{2}{\pi R} \iint \frac{x\, dx\, dy}{\sqrt{R^2 - y^2 - x^2}}$$

$$= -\frac{2}{\pi R} \int \sqrt{(R^2 - y^2 - x^2)}\, dy\, \Big]_{x=0}^{x=\sqrt{R^2-y^2}}$$

$$= \frac{2}{\pi R} \int \sqrt{(R^2 - y^2)}\, dy$$

$$= \frac{1}{\pi R} \left[y\sqrt{R^2 - y^2} + R^2 \sin^{-1} \frac{y}{R} \right]_0^R$$

$$= \tfrac{1}{2} R.$$

Similarly, $\bar{y} = \tfrac{1}{2} R = \bar{z}$.

This problem may be easily solved by the aid of elementary geometry. Conceive that the surface is divided into an indefinite number of small zones by equidistant planes which are perpendicular to the axis of x, in which case the area of the zones will be equal to each other. Conceive that these zones are reduced to the axis of x; they will then be uniformly distributed along that axis, and hence the centre of gravity will be $\frac{1}{2}R$ from the centre; and as the surface is symmetrical in respect to each of the three axes, we get the same result in respect to each.

73. *Centre of gravity of volumes, or heavy bodies.*
We have
$$dV = dx\,dy\,dz;$$
$$\therefore \bar{x}\iiint D\,dx\,dy\,dz = \iiint Dx\,dx\,dy\,dz; \qquad (79)$$
and similarly for $\bar{y}$ and $\bar{z}$.

If x is an axis of symmetry, (79) is sufficient.

If the surface is referred to polar coördinates, let

$\phi = AOx,$
$\theta = dOA,$
$\rho = Og,$

then,

$gd = d\rho,$
$gf = \rho d\theta,$
$gh = \rho \cos\theta d\phi,$

and,

$dV = \rho^2 d\rho \cos\theta d\theta d\phi,$

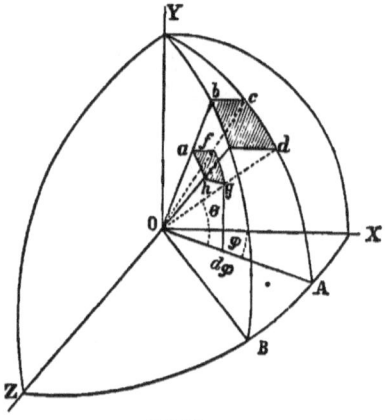

Fig. 45.

also,

$x = \rho \cos\theta \cos\phi; \; y = \rho \sin\theta; \text{ and } z = \rho \cos\theta \sin\phi.$

Hence, for a homogeneous body, we have

$$\begin{aligned} V\bar{x} &= \iiint \rho^3 \cos^2\theta \cos\phi\,d\rho\,d\theta\,d\phi; \\ V\bar{y} &= \iiint \rho^3 \cos\theta \sin\theta\,d\rho\,d\theta\,d\phi; \\ V\bar{z} &= \iiint \rho^3 \cos^2\theta \sin\phi\,d\rho\,d\theta\,d\phi. \end{aligned} \qquad (80)$$

If the volume be one of revolution about the axis of x, we have

$$dV = \int \pi y^2 dx;$$
$$V\bar{x} = \pi \int y^2 x \, dx.$$
(81)

EXAMPLE.

Find the centre of gravity of one-eighth of the volume of a homogeneous ellipsoid, contained within the three *principal* planes.

Let the equation of the ellipsoid be

$$\frac{x^2}{a^2} + \frac{y^2}{b^2} + \frac{z^2}{c^2} - 1 = 0;$$

then, equation (79) gives,

$$\bar{x} \int_0^c \int_0^Y \int_0^X dx \, dy \, dz = \int_0^c \int_0^Y \int_0^X x \, dx \, dy \, dz.$$

Performing the integration, we have

$$\tfrac{1}{6} \pi abc \, \bar{x} = \tfrac{1}{16} \pi a^2 bc;$$

$$\therefore \bar{x} = \tfrac{3}{8} a.$$

Similarly, $\bar{y} = \tfrac{3}{8} b$, and $\bar{z} = \tfrac{3}{8} c$.

Performing the above integration in the order of the letters x, y and z, and using the limits in the reverse order as indicated, we have for the *x-limits*,

$$x = 0, \text{ and } x = a\sqrt{1 - \frac{y^2}{b^2} - \frac{z^2}{c^2}} = X,$$

and the limits for y will be those which correspond to these values of x, or

for $x = 0$, $y = b\sqrt{1 - \frac{z^2}{c^2}} = Y$; and for $x = a\sqrt{1 - \frac{z^2}{c^2}}$, $y = 0$.

For the *first* member of the equation, we have

$$\int_0^c \int_0^Y \int_0^X dx \, dy \, dz = \int_0^c \int_0^Y x \, dy \, dz = \int_0^c \int_0^Y a\sqrt{1 - \frac{y^2}{b^2} - \frac{z^2}{c^2}} \, dy \, dz.$$

Consider $\sqrt{1 - \frac{z^2}{c^2}} = B$, a constant in reference to the *y-integration*, and we have

$$\int_0^c \tfrac{1}{4}a \left[y\sqrt{B^2 - \tfrac{y^2}{b^2}} + bB^2 \sin^{-1} \tfrac{y}{bB} \right]_0^{\tfrac{b}{c}\sqrt{c^2-z^2}} dz$$

$$= \tfrac{1}{4}\pi ab \int_0^c \left(1 - \tfrac{z^2}{c^2}\right) dz = \tfrac{1}{6}\pi abc.$$

For the *second* member, we have

$$\int_0^c \int_0^Y \int_0^X x\, dx\, dy\, dz = \int_0^c \int_0^Y \tfrac{1}{2}x^2\, dy\, dz \Big]_0^X = \tfrac{1}{2}a^2 \int_0^c \int_0^Y \left(1 - \tfrac{y^2}{b^2} - \tfrac{z^2}{c^2}\right) dy\, dz.$$

Consider z as constant in performing the *y-integration*, and we have

$$\tfrac{1}{2}a^2 \int_0^c \left(y - \tfrac{y^3}{3b^2} - \tfrac{z^2 y}{c^2} \right) dz \Big]_0^Y$$

$$= \tfrac{1}{2}a^2 \int_0^c \left[\tfrac{b}{c}(c^2 - z^2)^{\tfrac{1}{2}} - \tfrac{b}{3c^3}(c^2 - z^2)^{\tfrac{3}{2}} - \tfrac{b}{c^3} z^2 (c^2 - z^2)^{\tfrac{1}{2}} \right] dz.$$

$$= \tfrac{1}{16}\pi a^2 bc$$

$$\therefore \bar{x} = \frac{\tfrac{1}{16}\pi a^2 bc}{\tfrac{1}{6}\pi abc} = \tfrac{3}{8}a;$$

as given above.

74. When the centre of gravity of a body is known and the centre of gravity of a part is also known, the centre of the remaining part may be found as follows:—

Let $W =$ the weight of the whole body;
$\bar{x} =$ the distance from the origin to the centre of gravity of the body;
$w_1 =$ the weight of one part;
$x_1 =$ the distance of the centre of w_1 from the same origin;
$w_2 =$ the weight of the other part; and
$x_2 =$ the distance of the centre of w_2 from the same origin;

then
$$w_1 x_1 + w_2 x_2 = (w_1 + w_2)\bar{x} = W\bar{x};$$
and hence,
$$x_2 = \frac{W\bar{x} - w_1 x_1}{w_2}. \tag{32}$$

THEOREMS OF PAPPUS.

If the body is homogeneous, the volumes may be substituted for the weight.

EXAMPLE.

Let ABC be a cone in which the line BE joins the vertex and the centre of gravity of the base; and the cone ADC, having its apex D, on the line BE, and the same base AC be taken from the former cone, required the centre of gravity of the remaining part, $ADCB$.

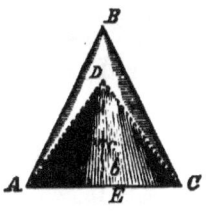

Fig. 46.

Let V = the volume of ACB,
$a = BE$,
$a_1 = DE$,
$\bar{x} = Ec = \frac{1}{4}a$ = the distance of the centre of ABC from E,
$x_1 = Eb = \frac{1}{4}a_1$ = the distance of the centre of ADC from E;

then,

$$\frac{a_1^3}{a^3} V = \text{the volume of } ACD,$$

and (82) becomes

$$x_2 = \frac{V \cdot \frac{1}{4}a - V \cdot \frac{a_1^3}{a^3} \cdot \frac{1}{4}a_1}{V - V \cdot \frac{a_1^3}{a^3}}$$

$$= \frac{1}{4}\frac{a^4 - a_1^4}{a^3 - a_1^3}.$$

CENTROBARIC METHOD, OR

75. THEOREMS OF PAPPUS OR OF GULDINUS.

Multiply both members of the second of (74) by 2π, and it may be reduced to

$$2\pi \bar{y} s = 2\pi \int y \, ds, \qquad (83)$$

the second member of which is the area generated by the revolution of a line whose length is s about the axis of x, and the first member is the circumference described by the centre of gravity of the line, multiplied by the length of the line; hence, *the area generated by the revolution of a line about a fixed axis equals the length of the line multiplied by the circumference described by the centre of gravity of the line.*

This is one of the theorems, and the following is the other. From the second of (76), we find

$$2\pi \bar{y} A = \int \pi y^2 dx.$$

The right-hand member, integrated between the proper limits, is the volume generated by the revolution of a plane area about the axis of x. The plane area must lie wholly on one side of the axis. In the first member of the equation, A is the area of the plane curve, and $2\pi \bar{y}$ is the circumference described by its centre of gravity. Hence, *the volume generated by the revolution of a plane curve which lies wholly on one side of the axis, equals the area of the curve multiplied by the circumference described by its centre of gravity.*

EXAMPLES.

1. Find the surface of a ring generated by the revolution of a circle, whose radius is r, about an axis whose distance from the centre is c.

Ans. $4\pi^2 rc.$

2. The surface of a sphere is $4\pi r^2$, and the length of a semicircumference is πr; required the ordinate to the centre of gravity of the arc of a semicircle.

3. Required the volume generated by an ellipse, whose semi-axes are a and b, about an axis of revolution whose distance from the centre is c; c being greater than a or b.

(Observe that the volume will be the same for all positions of the axes a and b in reference to the axis of revolution.)

4. The volume of a sphere is $\frac{4}{3}\pi r^3$, and the area of a semicircle is $\frac{1}{2}\pi r^2$; show that the ordinate to the centre of gravity of the semicircle is $\bar{y} = \dfrac{4r}{3\pi}.$

76. ADDITIONAL EXAMPLES.

1. Find the centre of gravity of the quadrant of the circumference of a circle contained between the axes x and y, the origin being at the centre.

$$\text{Ans. } \bar{x} = \frac{2r}{\pi} = \bar{y}.$$

2. Find the distance of the centre of gravity of the arc of a cycloid from the vertex, r being the radius of the generating circle.

$$\text{Ans. } \bar{y} = \tfrac{4}{3}r.$$

3. Find the centre of gravity of one-half of the loop of a leminiscate, of which the equation is $r^2 = 2a \cos 2\theta$, l being the length of the half loop.

$$\text{Ans. } \bar{x} = \frac{a^2}{\sqrt{2}\,l}; \quad \bar{y} = \frac{a^2}{l} \frac{\sqrt{2}-1}{\sqrt{2}}.$$

4. Find the centre of gravity of the helix whose equations are
$$x = a \cos \phi, \; y = a \sin \phi, \; z = na\phi,$$
the helix starting on the axis of $\bar{x}$.

$$\text{Ans. } \bar{x} = na\frac{y}{z}; \quad \bar{y} = na\frac{a-x}{z}; \text{ and } \bar{z} = \tfrac{1}{2}z.$$

5. Find the centre of gravity of the perimeter of a triangle in space.

6. If x_0 and y_0 are initial points of a curve, find the curve such that $m\bar{x} = x - x_0$, and $n\bar{y} = y - y_0$.

7. A curve of given length joins two fixed points; required its form so that its centre of gravity shall be the lowest possible.
(This may be solved by the Calculus of Variations).

Ans. A Catenary.

8. Find the centre of gravity of a trapezoid.
Let $ADEB$ be the trapezoid, in which DE and AB are the parallel sides. Produce AD and BE until they meet in C, and join C with F, the middle point of the base; then the centre of gravity will be at some point g on this line. The centre of the triangle ACB will be on CF, and at a distance of $\tfrac{1}{3}CF$ from F; and similarly that of DCE will be on the same line, and at a distance of $\tfrac{1}{3}CG$ from G; then, by (79) we may find Fg.

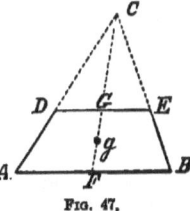

Fig. 47.

$$\text{Ans. } Fg = \tfrac{1}{3}FG\frac{AB + 2DE}{AB + DE}.$$

(If DE is zero, we have $Fg = \tfrac{1}{3}FC$ for the centre of gravity of the triangle ABC.)

9. Find the centre of gravity of the quadrant of an ellipse, whose equation is $a^2y^2 + b^2x^2 = a^2b^2$.

$$\text{Ans. } \bar{x} = \tfrac{4}{3}\frac{a}{\pi}; \quad \bar{y} = \tfrac{4}{3}\frac{b}{\pi}.$$

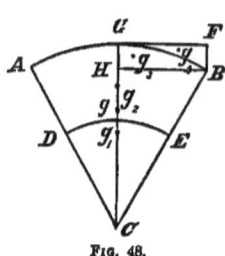

Fig. 48.

10. Find the centre of gravity of the circular sector ABC.

Let the angle $ACB = 2\theta$; then

$$\bar{x} = Cg = \tfrac{2}{3} \frac{r \sin \theta}{\theta}.$$

11. Find the centre of gravity of a part of a circular annulus $ABED$.

Let $AC = r$, $DC = r_1$, and $ACB = 2\theta$; then

$$\text{Ans. } Cg_2 = \tfrac{2}{3} \cdot \frac{\sin \theta}{\theta} \cdot \frac{r^2 + rr_1 + r_1^2}{r + r_1}.$$

12. Find the centre of gravity of the circular spandril FGB.

13. Find the centre of gravity of a circular segment.

$$\text{Ans. Dist. from } C = \frac{(\text{chord})^3}{12 \text{ area of segment}}.$$

14. Find the distance of the centre of gravity of a complete cycloid from its vertex, r being the radius of the generating circle.

$$\text{Ans. } \bar{y} = \tfrac{5}{6} r.$$

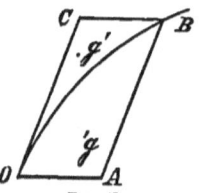

Fig. 49.

15. Find the centre of gravity of the parabolic spandril OCB, Fig. 49, in which $OC = y$, and $CB = x$.

$$\text{Ans. } \bar{x} = \tfrac{3}{10} x; \\ \bar{y} = \tfrac{3}{4} y.$$

16. Find the centre of gravity of a loop of the lemniscate, whose equation is $r^2 = a^2 \cos 2\theta$.

$$\text{Ans. } \bar{x} = \frac{\pi}{2^{\frac{5}{2}}} a.$$

17. Find the centre of gravity of a hemispherical surface.

$$\text{Ans. } \bar{x} = \tfrac{1}{2} r.$$

18. Find the centre of gravity of the surface generated by the revolution of a semi-cycloid about its base, a being the radius of the generating circle and $\bar{x}$ the distance from the base.

$$\text{Ans. } \bar{x} = \tfrac{8}{15} a.$$

19. The centre of gravity of the volume of a paraboloid of revolution is

$$\bar{x} = \tfrac{2}{3} x.$$

20. The centre of gravity of one half of an ellipsoid of revolution, of which the equation is $a^2 y^2 + b^2 x^2 = a^2 b^2$, is

$$\bar{x} = \tfrac{3}{8} a.$$

21. The centre of gravity of a rectangular wedge is

$$\bar{x} = \tfrac{3}{8} a.$$

22. The centre of gravity of a semicircular cylindrical wedge, whose radius is r, is

$$\bar{x} = \tfrac{3}{16}\pi r.$$

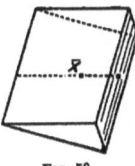

Fig. 50.   Fig. 51.

23. The vertex of a right circular cone is in the surface of a sphere, the axis of the cone passing through the centre of the sphere, the base of the cone being a portion of the surface of the sphere. If 2θ be the vertical angle of the cone, required the distance of the centre of gravity from the apex.

$$Ans. \quad \frac{1 - \cos^5\theta}{1 - \cos^4\theta} r.$$

24. Find the distance from G, Fig. 48, to the centre of gravity of a spherical sector generated by the revolution of a circular sector GCA, about the axis GC.

$$Ans. \quad \tfrac{1}{4}(GC + \tfrac{1}{2}GH).$$

25. A circular hole with a radius r is cut from a circular disc whose radius is R; required the centre of gravity of the remaining part, when the hole is tangent to the circumference of the disc.

26. Find the centre of gravity of the frustum of a pyramid or cone.

It will be in the line which joins the centre of gravity of the upper and lower bases. Let h be the length of this line, and a and b be corresponding lines in the lower and upper bases respectively, required the distance, measured on the line h, of the centre from the lower end.

$$Ans. \quad \bar{x} = \tfrac{1}{4}h \, \frac{a^2 + 2ab + 3b^2}{a^2 + ab + b^2}.$$

If $b = 0$, we have the distance of the centre of a pyramid or cone from the base equal to $\tfrac{1}{4}h$.

27. Find the centre of gravity of the octant of a sphere in which the density varies directly as the nth power of the distance from the centre, r being the radius of the sphere.

$$Ans. \quad \bar{x} = \frac{n+3}{2n+8}r = \bar{y} = \bar{z}.$$

28. Find the centre of gravity of a paraboloid of revolution of uniform density whose axis is a.

$$Ans. \quad \bar{x} = \tfrac{2}{3}a.$$

SOME GENERAL PROPERTIES OF THE CENTRE OF GRAVITY.

77. *When a body is at rest on a surface, a vertical through the centre of gravity will fall within the support.*

For, if it passes without the support, the reaction of the surface upward and of the weight downward form a statical couple, and rotation will result.

78. *When a body is suspended at a point, and is at rest, the centre of gravity will be vertically under the point of suspension.*

The proof is similar to the preceding. When the preceding conditions are fulfilled the body is in equilibrium.

79. A body is in a condition of *stable equilibrium* when, if its position be slightly disturbed, it tends to return to its former position when the disturbing force is removed; of *unstable equilibrium* if it tends to depart further from its position of rest when the disturbing force is removed; and of *indifferent equilibrium* if it remains at rest when the disturbing force is removed.

EXAMPLE.

A paraboloid of revolution rests on a horizontal plane; required the inclination of its axis.

Let P be the point of contact of the paraboloid and plane, then will the vertical through P pass through the centre of gravity G, and PG will be a normal to the paraboloid.

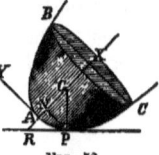

Fig. 52.

The equation of a vertical section through the centre is $y^2 = 2px$, in which x is the axis, the origin being at the vertex.

Let $a = AX =$ the altitude of the paraboloid;
$\theta = GRP =$ the inclination of the axis;
then, $AG = \tfrac{2}{3} a$, (see example 28 on the preceding page);
$AN = \tfrac{2}{3} a - p$;
hence,

$$\tan \theta = \frac{dy}{dx} = \frac{p}{y} = \sqrt{\frac{p}{2x}} = \sqrt{\frac{p}{2(\tfrac{2}{3}a - p)}}$$

which will be positive and real as long as $\tfrac{2}{3} a$ exceeds p. In this case the equilibrium is *stable*. When $\frac{2}{3}a$ exceeds p it will also rest on the apex, but the equilibrium for this position is *unstable*. When $\frac{2}{3}a = p$, $\theta = 90°$, and the segment will rest only on the apex. When $\frac{2}{3}a$ is less than p, tan θ becomes imaginary, and hence, this analysis fails to give the position of rest; but by independent reasoning we find, as before, that it will rest on the apex, and that the equilibrium will be *stable*.

80. *In a plane material section the sum of the products found by multiplying each elementary mass by the square of its distance from an axis, equals the sum of the similar products in reference to a parallel axis passing through the centre, plus the mass multiplied by the square of the distance between the axes.*

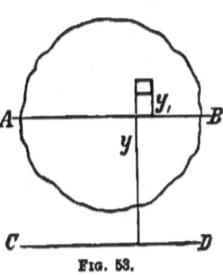

Fig. 53.

Let AB be an axis through the centre,
 CD a parallel axis,
 $D =$ the distance between AB and CD,
 $dm =$ an elementary mass,
 $y_1 =$ the ordinate from AB to m,
 $y =$ the ordinate from CD to m, and
 $M =$ the mass of the section.

Then
$$y^2 = (y_1 + D)^2 = y_1^2 + 2y_1 D + D^2.$$

Multiply by dm and integrate, and we have
$$\int y^2 dm = \int y_1^2 dm + 2D \int y_1 dm + D^2 \int dm.$$

But since AB passes through the centre, the integral of $y_1 dm$, when the whole section is included, is zero (see Eq. 71b), and $\int dm = M$; hence,
$$\int y^2 dm = \int y_1^2 dm + MD^2. \qquad (83)$$

Similarly, if dA be an elementary area, and A the total area, we have
$$\int y^2 dA = \int y_1^2 dA + AD^2.$$

81. *In any plane area, the sum of the products of each elementary area multiplied by the square of its distance from an axis, is least when the axis passes through the centre.*

This follows directly from the preceding equation, in which the first member is a minimum for $D = 0$.

CENTRE OF THE MASS.

82. *The centre of the mass is such a point that, if the whole mass be multiplied by its distance from an axis, it will equal the sum of the products found by multiplying each elementary mass by its distance from the same axis.*

Let m = an elementary mass;
M = the total mass;
$x_1, y_1,$ and z_1 be the respective coördinates, of the centre of the mass, and
$x, y,$ and z the general coördinates,
then, according to the definition, we have

$$\left. \begin{array}{l} Mx_1 = \Sigma mx; \\ My_1 = \Sigma my; \\ Mz_1 = \Sigma mz; \end{array} \right\} \quad (84)$$

which being the same as (70) shows that when we consider the force of gravity as constant for all the particles of a body, the centre of the mass coincides with the centre of gravity. This is *practically* true for finite bodies on the surface of the earth, although the centre of gravity is *actually* nearer the earth than the centre of the mass is.

If the origin of coördinates be at the centre of the mass, we have

$$\Sigma mx = 0; \quad \Sigma my = 0; \quad \Sigma mz = 0; \quad (84a)$$

which are the same as (71a).

CHAPTER IV.

NON-CONCURRENT FORCES.

83. EQUILIBRIUM OF A SOLID BODY ACTED UPON BY ANY NUMBER OF FORCES APPLIED AT DIFFERENT POINTS AND ACTING IN DIFFERENT DIRECTIONS.

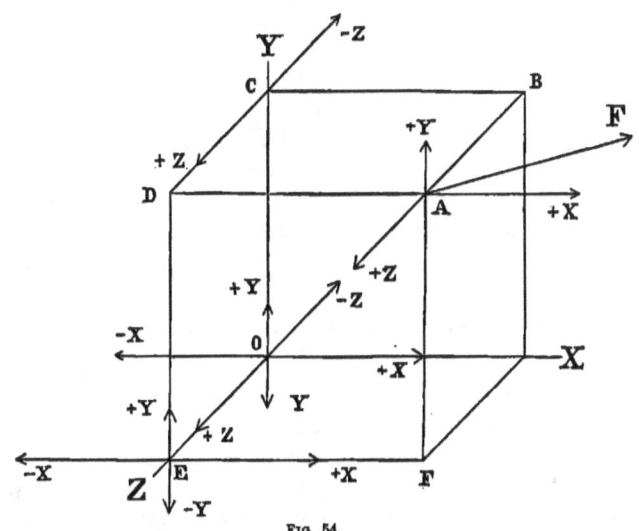

FIG. 54.

Let A be any point of a body, at which a force F is applied, and O the origin of coördinates, which, being chosen arbitrarily, may be within or without the body. On the coördinate axes construct a parallelopipedon having one of its angles at O, and the diagonally opposite one at A.

Let the *typical force* F be in the first angle and acting away from the origin, so that all of its direction-cosines will be positive; then will the sign of the axial component of any force be the same as that of the trigonometrical cosine of the angle which the direction of the force makes with the axis.

Let a = the angle between F and the axis of x,
$\beta =$ "   "   "   "   "   "   "   " y,
$\gamma =$ "   "   "   "   "   "   "   " z,
then will the X, Y, and Z-*components* of the force F be
$$X = F\cos a,$$
$$Y = F\cos \beta,$$
$$Z = F\cos \gamma.$$

The point of application of the X-*component*, being at any point in its line of action, may be considered as at D, where its action-line meets the plane yz. At E introduce two equal and opposite forces, each equal and parallel to X, and since they will equilibrate each other, the mechanical effect of the system will be the same as before they were introduced. Combining the force $+ X$ at D with $- X$ at E, we have a *couple* whose arm is $DE = y =$ the *y-ordinate* of the point A. This couple, according to Article 54, will be negative, hence, its moment is
$$- Xy.$$

Hence, a force $+ X$ at A produces the same effect upon a body as the couple $- Xy$, and a force $+ X$ at E.

At the origin O introduce two equal and opposite forces, each equal to X, acting along the axis of x. This will not change the mechanical effect of the system. Combining $- X$ at O with $+ X$ at E, we have the couple $+ Xz$, and the force $+ X$ remaining at O. Hence, *a single force $+X$ at A is equivalent to an equal parallel force at the origin of coördinates, and the two couples,*
$$- Xy \text{ and } + Xz.$$

Treating the Y-*component* in a similar manner, we have the force
$$+ Y \text{ at the origin, and}$$
the moments,
$$+ Yx \text{ and } - Yz;$$
and similarly for the Z-*component*, the force
$$+ Z \text{ at the origin, and}$$
the moments,
$$- Zx \text{ and } + Zy.$$

But the couples $+ Zy$ and $- Yz$, have the common axis x, and hence are equivalent to a single couple which is equal to the algebraic sum of the two; and similarly for the others; hence, the *six* couples may be reduced to the *three* following:

$Zy - Yz$, having x for an axis;
$Xz - Zx$, " y "   "   ";
$Yx - Xy$, " z "   "   ";

hence, *for the single force F acting at A there may be substituted the three axial components of the force acting at the origin of coördinates, and three pairs of couples having for their axes the respective coördinate axes.*

If there be a *system* of forces, in which
F_1, F_2, F_3, etc., are the forces,
x_1, y_1, z_1, the coördinates of the point of application of F_1,
x_2, y_2, z_2, "   "   "   "   "   "   "   " F_2,
etc., etc., etc.,
a_1, a_2, a_3, etc., the angles made by F_1, F_2, etc., respectively with the axis of x,
β_1, β_2, β_3, etc., the angles made by the forces with y, and
γ_1, γ_2, γ_3, etc., the corresponding angles with z;
then resolving each of the forces in the same manner as above, we have the *axial components*

$$\left. \begin{array}{l} X = F_1 \cos a_1 + F_2 \cos a_2 + F_3 \cos a_3 + \text{etc.} = \Sigma F \cos a\, ; \\ Y = F_1 \cos \beta_1 + F_2 \cos \beta_2 + F_3 \cos \beta_3 + \text{etc.} = \Sigma F \cos \beta\, ; \\ Z = F_1 \cos \gamma_1 + F_2 \cos \gamma_2 + F_3 \cos \gamma_3 + \text{etc.} = \Sigma F \cos \gamma\, ; \end{array} \right\} \quad (85)$$

and the *component moments*

$$\left. \begin{array}{l} Zy - Yz = \Sigma(Fy \cos \gamma - Fz \cos \beta) = L\, ; \\ Xz - Zx = \Sigma(Fz \cos a - Fx \cos \gamma) = M\, ; \\ Yx - Xy = \Sigma(Fx \cos \beta - Fy \cos a) = N\, ; \end{array} \right\} \quad (86)$$

in which L, M, and N are used for brevity.

RESULTANT FORCE AND RESULTANT COUPLE.

84. Let $R =$ the resultant of a system of forces concurring at the origin of coördinates, and having the same magnitudes and directions as those of the given forces;

a, b, and c = the angles which it makes with the axes x, y, and z respectively;

G = the moment of the resultant couple;

d, e, and f = the angles which the axis of the resultant couple makes with the axes x, y, and z respectively;

then

$$\left.\begin{aligned} X &= R \cos a\,; \\ Y &= R \cos b\,; \\ Z &= R \cos c\,; \end{aligned}\right\} \qquad (87)$$

$$\left.\begin{aligned} L &= G \cos d\,; \\ M &= G \cos e\,; \\ N &= G \cos f. \end{aligned}\right\} \qquad (88)$$

If a force and a couple, equal and opposite respectively to the resultant force and resultant couple, be introduced into the system, there will be equilibrium, and R and G will both be zero. Hence, for equilibrium, we have

$$X = 0\,; \qquad Y = 0\,; \qquad Z = 0\,; \qquad (89)$$

$$L = 0\,; \qquad M = 0\,; \qquad N = 0. \qquad (90)$$

85. DISCUSSION OF EQUATIONS (87) AND (88).

1. *Suppose that the body is perfectly free to move in any manner.*

   a. If the forces concur and are in equilibrium, equations (87) only are necessary, and are the same as equations (60); hence, we will have

   $$X = 0, \quad Y = 0, \quad Z = 0.$$

   b. If $R = 0$ and G is finite, equations (88) only are necessary.

   c. If R and G are both finite, then all of equations (87) and (88) may be necessary.

2. If *one point* of the body is *fixed*, there can be no translation, and equations (88) will be sufficient.

3. If an *axis parallel to x is fixed* in the body, there may be translation along that axis, and rotation about it; hence, the 1st of (87) and the 1st of (88) are sufficient.

4. If *two points are fixed*, it cannot translate, but may rotate; and by taking x so as to pass through the two points, the equation $L = 0$ is sufficient.

5. If *one point only* is confined to the *plane xy*, the body will have every degree of freedom except moving parallel to z, and hence, all of equations (87) and (88) are necessary except the 3d of (87).

6. If *three points*, not in the same straight line, are confined to the *plane xy*, it may rotate about z, but cannot move parallel to z; hence, the 1st and 2d of (87) and the 3d of (88) are necessary and sufficient.

7. If *two axes parallel to* x are fixed, the body can move only parallel to x, and the 1st of (87) is sufficient.

8. If the *forces* are parallel to the axis of y, there can be translation parallel to y only, and rotation about x and z.

9. If the forces are in the plane xy, the equations for equilibrium become

$$\left. \begin{array}{l} X = \Sigma F \cos a = R \cos a = 0; \\ Y = \Sigma F \cos \beta = R \cos b = 0; \\ Yx - Xy = \Sigma (Fx \cos \beta - Fy \cos a) = 0. \end{array} \right\} \quad (91)$$

[OBS. In a mechanical sense, whatever holds a body is a force. Hence, when we say "a point is fixed," or, "an axis is fixed," it is equivalent to introducing an indefinitely large resisting force. Instead of finding the value of the resistance, it has, in the preceding discussion, been eliminated. When we say "the body cannot translate," it is equivalent to saying that finite active forces cannot overcome an infinite resistance.]

86. Applications of Equations (91).

a. PROBLEMS IN WHICH THE TENSION OF A STRING IS INVOLVED.

1. *A body AB, whose weight is W, rests at its lower end upon a perfectly smooth horizontal plane, and at its upper end against a perfectly smooth vertical plane: the lower end is prevented from sliding by a string CB. Determine the tension on the string, and the pressure upon the horizontal and vertical planes.*

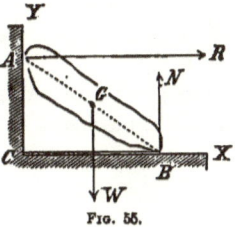

Fig. 55.

Take the origin of coördinates at C, the axis of x coinciding with CB, and y with AC, x being positive to the right, and y positive upwards.

Let $W =$ the weight of the body whose centre of gravity is at G;

$R =$ the reaction of the vertical wall, and, since there is no friction, its direction will be perpendicular to AC;

$N =$ the reaction of the horizontal plane, which will be perpendicular to CB;

$l =$ the horizontal distance from C to the vertical through the centre of gravity;

$t =$ the tension of the string;

then equations (91) become

$$X = R + t = 0;$$
$$Y = N + W = 0;$$
$$Xy - Yx = R.AC + t.0 + N.CB + W.l = 0;$$

in which all the quantities are treated as positive.

Solving these equations, gives

$$N = -W, \text{ and}$$
$$R = \frac{CB - l}{AC}W = -t.$$

If the centre of gravity is at the middle of AB, then

$$R = \frac{CB}{2AC}\ W = \tfrac{1}{2}W \tan BAC.$$

We thus see that the horizontal plane sustains the whole weight, and that the tension of the string is equal and opposite to the reaction of the vertical plane.

The reaction N and the weight W, being equal and parallel, and acting in opposite directions, constitute a couple whose *arm* is equal to $CB - l$. Similarly the tension t and the reaction R constitute another couple whose *arm* is AC; and since there is equilibrium in reference to rotation, we have

$$W(CB - l) = R.AC;$$
$$\therefore R = \frac{CB - l}{AC}\ W,$$

as before.

The direction of the forces will generally be known from the conditions of the problem, and it is generally best to enter them in the equations with their proper signs; but, as we have seen above, this is not always necessary. When only three forces are involved in a system their *relative* signs may be determined from the analysis.

[OBS. It will be a profitable exercise for the student to solve the same example by taking the origin of coördinates at different points.]

2. A ladder rests on a smooth horizontal plane and against a vertical wall, the lower end being held by a horizontal string; a person ascends the ladder, required the pressure against the wall for any position on the ladder.

3. A uniform beam, whose length is AB and weight W, is held in a horizontal position by the inclined string CD, and carries a weight P at the extremity; required the tension of the string.

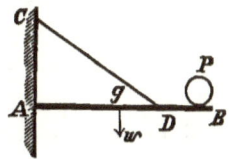

Fig. 56.

$$\text{Ans. } t = \frac{AB}{AD} \cdot \frac{DC}{AC} (P + \tfrac{1}{2}W).$$

4. A prismatic piece AB is permitted to turn freely about the lower end A, and is held by a string CE; given the position of the centre of gravity, the weight W of the piece, the inclination of the piece and string, and the point of attachment E; required the tension of the string, and the pressure against the lower end of the beam at A.

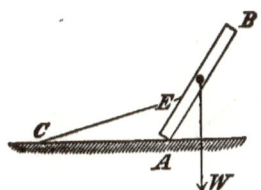

Fig. 57.

5. A heavy piece AB is supported by two cords which pass over pulleys C and D, and have weights P_1 and P attached to them; required the inclination to the horizontal of the line AB joining the points of attachment of the cord.

(Consider the pulleys as reduced to the points C and D.)

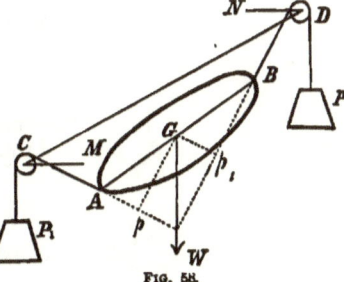

Fig. 58.

Let G, the centre of gravity of AB, be on the line joining the points of attachment A and B;
$a = AG$; $b = BG$;
$i =$ the angle DCM;
$\delta =$ the inclination of BA to DC;
$\alpha = DCA$; and $\beta = CDB$.

Resolving horizontally and vertically, we have
$X = - P_1 \cos MCA + P \cos NDB + W \cos 90° = 0$;
$\quad = - P_1 \cos (\alpha - i) + P \cos (\beta + i) = 0$; $\quad (a)$
$Y = P_1 \sin (\alpha - i) + P \sin (\beta + i) - W = 0.$ $\quad (b)$

Taking the origin of moments at G, making Gp perpendicular to AC, and Gp_1 perpendicular to DB, we have
$- P_1 \times Gp + P \times Gp_1 + W \times 0 = 0$;
or $\quad - P_1 \cdot a \sin (\alpha + \delta) + P \cdot b \sin (\beta - \delta) = 0.$ $\quad (c)$

The angle i is given by the conditions of the problem; hence the three equations (a), (b), and (c) are sufficient to determine the angles α, β, and δ, when the numerical values of the given quantities are known. The inclination will be $\delta + i$.

6. Suppose, in Fig. 58, that the lengths of the strings AC and BD are given, required the inclination of AB.

[The solution of this problem involves an equation of the 8th degree].

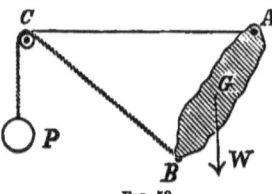
Fig. 59.

7. A heavy piece AB, Fig. 59, is free to swing about one end A, and is supported by a string BC which passes over a pulley at C, and is attached to a weight P; find the angle ACB when they are in equilibrium.

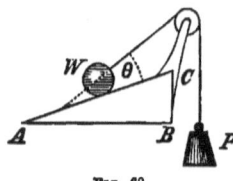
Fig. 60.

8. A weight W rests on a plane whose inclination to the horizontal is i, and is held by a string whose inclination to the plane is θ; required the relation between the tension P and the weight, and the value of the normal pressure upon the plane.

Ans. $P = \dfrac{\sin i}{\cos \theta} W$; $\quad$ Normal pressure $= \dfrac{\cos (\theta + i)}{\cos \theta} W.$

b. EQUILIBRIUM OF PERFECTLY SMOOTH BODIES IN CONTACT WITH EACH OTHER.

9. *A heavy beam rests on two smooth inclined planes, as in Fig. 61; required the inclination of the beam to the horizontal, and the reactions of the respective planes.*

Let AC and CB be the inclined planes; AB the beam whose centre of gravity is at G. When it *rests*, the reactions of the planes must be normal to the planes, for otherwise they would have a component parallel to the planes which would produce motion.

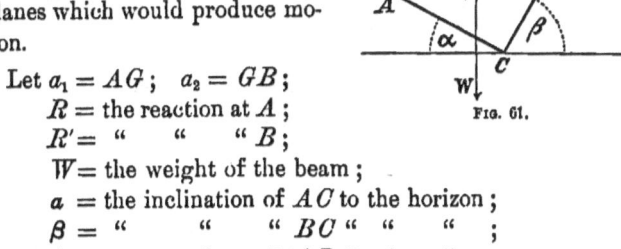

Fig. 61.

Let $a_1 = AG$; $a_2 = GB$;
$R =$ the reaction at A ;
$R' =$ " " " B ;
$W =$ the weight of the beam ;
$a =$ the inclination of AC to the horizon ;
$\beta =$ " " " BC " " " ;
$\theta =$ " " " AB " " " .

Take the origin of coördinates at the centre of gravity G of the body, x horizontal and y vertical.

The forces resolved horizontally give

$$X = R \sin a - R' \sin \beta + W \cos 90° = 0 ; \qquad (a)$$

and vertically,

$$Y = R \cos a + R' \cos \beta - W \sin 90° = 0. \qquad (b)$$

The moment of $R \sin a$ is, ... $+ R \sin a \times AG \sin \theta$.
" " " $R' \sin \beta$ is, ... $+ R' \sin \beta \times GB \sin \theta$.
" " " $W \cos 90°$ is, ... 0.
" " " $R \cos a$ is, ... $- R \cos a \times AG \cos \theta$.
" " " $R' \cos \beta$ is, ... $+ R' \cos \beta \times GB \cos \theta$.

Hence $Xy - Yx = Ra_1 \sin a \sin \theta + R'a_2 \sin \beta \sin \theta$
$\qquad - Ra_1 \cos a \cos \theta + R'a_2 \cos \beta \cos \theta$
$\qquad = - Ra_1 \cos(a + \theta) + R'a_2 \cos(\beta - \theta) = 0. \qquad (c)$

It is generally better to deduce the values of the moments directly from the definitions; (see Articles 51 to 57). To do this in the present case, let fall from G the perpendiculars aG and bG upon the action-lines of the respective forces; then

$$bG = a_2 \sin (90° - (\beta - \theta)) = a_2 \cos (\beta - \theta);$$
$$aG = a_1 \sin (90° - (\alpha + \theta)) = a_1 \cos (\alpha + \theta);$$

and we have

the moments $= -R \cdot aG + R' \cdot bG = -Ra_1 \cos (\alpha + \theta) + R'a_2 \cos (\beta - \theta) = 0$; as given above.

Solving equations (a), (b), and (c), we find

$$R = \frac{\sin \beta}{\sin (\alpha + \beta)} W; \qquad R' = \frac{\sin \alpha}{\sin (\alpha + \beta)} W;$$

$$\tan \theta = \frac{a_1 \cos \alpha \sin \beta - a_2 \sin \alpha \cos \beta}{(a_1 + a_2) \sin \alpha \sin \beta}.$$

If $R = R'$, then
$$\sin \beta = \sin \alpha;$$

which are the conditions necessary to make the normal reactions equal to each other.

The reactions prolonged will meet the vertical through the centre of gravity at a common point D, and if the beam be suspended at D by means of the two cords DA and DB it will retain its position when the planes AC and CB are removed.

If $\beta = 90°$, the plane CB will be vertical, and we find

$$R = W \sec \alpha; \qquad R' = \frac{\sin \alpha}{\cos \alpha} W = W \tan \alpha;$$

$$\tan \theta = \frac{a_1}{a_1 + a_2} \cot \alpha.$$

If $a_1 = a_2$, then
$$\tan \theta = \frac{\sin (\alpha - \beta)}{2 \sin \alpha \sin \beta}.$$

If $\beta = 90°$ and $a = 0°$, then

$$R' = 0, \quad \theta = 0, \quad \text{and } R = W.$$

A special case is that in which the beam coincides with one of the planes. The formulas do not apply to this case.

10. Two equal, smooth cylinders rest on two smooth planes whose inclinations are a and β respectively; required the inclination, θ, of the line joining their centres.

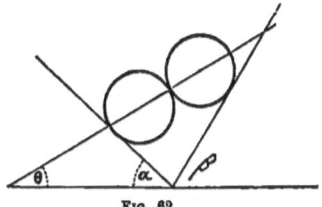

Fig. 62.

Ans. $\tan \theta = \tfrac{1}{2}(\cot a - \cot \beta)$.

11. A heavy, uniform, smooth beam *rests on* one edge of a box at C, and *against* the vertical side opposite; required its inclination to the vertical. Let g be the centre of gravity.

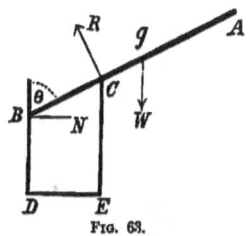

Ans. $\sin \theta = \sqrt[3]{\dfrac{DE}{Bg}}$.

Fig. 63.

12. Three equal, smooth cylinders are placed in a box, the two lower ones being tangent to the sides of the box and to each other, and the other placed above them and tangent to both; required the pressure against the bottom and sides of the box.

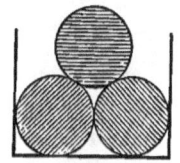

Fig. 64.

Ans. Pressure on the bottom = total weight of the cylinders.

Pressure on one side = $\tfrac{1}{2}$ *weight of one cylinder* $\times \tan 30°$.

13. Two homogeneous, smooth, prismatic bars rest on a horizontal plane, and are prevented from sliding upon it; required their position of equilibrium when leaning against each other.

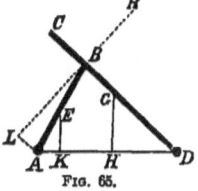

Fig. 65.

Let AB and CD be the two bars, resting against each other at B; then will they be in equilibrium when the resultant of their pressures at B is perpendicular to the face of CD.

Let $b = AB$; $c = CD$; $x = BD$;
$a = AD =$ the distance between the lower ends of the bars;
$W =$ the weight of AB;
$W_1 =$ the weight of CD;
E and G the respective centres of gravity of the bars, which will be at the middle of the pieces; then we have

$$2(a^2 + b^2 - x^2) x^3 W = c(a^2 - b^2 + x^2)(-a^2 + b^2 + x^2) W_1;$$

which is an equation of the fifth degree, and hence always admits of one real root.

14. *The upper end of a heavy piece rests against a smooth, vertical plane, and the lower end in a smooth, spherical bowl; required the position of equilibrium.*

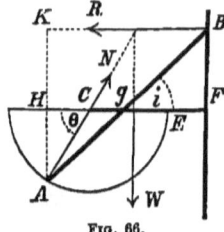
Fig. 66.

Let AB be the piece, BF the vertical surface, EA the spherical surface, and g the centre of gravity of the piece.
When it is in equilibrium, the reaction at the lower end will be in the direction of a normal to the surface, and hence will pass through C, the centre of the sphere, and the reaction of the vertical plane will be horizontal.

Let $W =$ the weight of the piece;
$r =$ the radius of the sphere;
$a = Ag$; $b = Bg$; $l = AB$; $d = CF$;
$R =$ the reaction of the vertical plane;
$N =$ the reaction of the spherical surface;
$i =$ the inclination of the beam to the horizontal;
$\theta =$ the inclination of the radius to the horizontal.

Take the origin of coördinates at g, x horizontal and y vertical; and we have

$$X = N \cos \theta - R = 0;$$
$$Y = N \sin \theta - W = 0;$$

$$\text{Moments} = + R.b \sin i - N.a \sin(\theta - i) = 0;$$

and the geometrical relations give,
$$l \cos i = KB = HF = d + r \cos \theta.$$
From these equations, we have
$$N = W \operatorname{cosec} \theta; \quad R = W \cot \theta;$$
$$a \sin (\theta - i) - b \cos \theta \sin i = 0,$$
which, by developing and reducing, becomes
$$(a + b) \tan i = a \tan \theta;$$
this, combined with the fourth equation above, will determine i and θ.

The position is independent of the weight of the piece, but depends upon the position of its centre of gravity.

15. A heavy prismatic bar of infinitesimal cross-section rests against the concave arc of a vertical parabola, and a pin placed at the focus; required the position of equilibrium.

Let $l = AB =$ length of the bar; $p = CD$ = one-fourth the parameter of the parabola, C being the focus, and $\theta = ACD$.

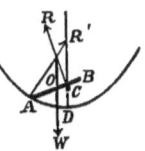

FIG. 67.

$$Ans.\ \theta = 2 \sin^{-1} \left(\frac{p}{l}\right)^{\frac{1}{4}}.$$

16. Required the form of the curve such that the bar will rest in all positions.

Ans. The polar equation is $r = \frac{1}{2}l + c \sec \theta$, in which l is the length of the bar, and c an arbitrary constant. It is the equation of the conchoid of Nicomedes.

C. INDETERMINATE PROBLEMS.

17. *To determine the pressures exerted by a door upon its hinges.*

Let $W =$ the weight of the door;

$a =$ the distance between the hinges;

$b =$ the horizontal distance from the centre of gravity of the door to the vertical line which passes through the hinges;

$F =$ the vertical reaction of the upper hinge;

$F_1 =$ the vertical reaction of the lower hinge;

$H =$ the horizontal reaction of the upper hinge;

$H_1 =$ the horizontal reaction of the lower hinge;

then

$$X = H - H_1 = 0;$$
$$Y = F + F_1 - W = 0;$$
$$Xy - Yx = Ha - Wb = 0;$$

which give

$$H = H_1 = \frac{b}{a} W; \text{ and}$$

$$F + F_1 = W.$$

The result, therefore, is indeterminate, but we can draw two general inferences: 1st, *The horizontal pressures upon the hinges are equal to each other but in opposite directions;* and, 2d, *The vertical reaction upon both hinges equals the weight of the door.*

It is necessary to have additional data in order to determine the *actual pressure* on each hinge. The ordinary imperfections of workmanship will cause one to sustain more weight than the other, but as they wear they may approach an equality.

The horizontal and vertical pressures being known, the actual pressures may be found by the *triangle of forces*. If the upper end sustains the whole weight, the total pressure upon it will be $\dfrac{W}{a}\sqrt{a^2 + b^2}$. If each sustains one-half the weight, the pressure on each will be one-half this amount.

18. A rectangular stool rests on four legs, one being at each corner of the stool; required the pressure on each.

(The data are insufficient.)

19. A weight P is supported by three unequally inclined struts; required the amount which each will sustain.

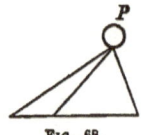

Fig. 68.

[OBS. If more conditions are given than there are quantities to be determined, they will either be redundant or conflicting.]

d. STRESS ON FRAMES.

20. *Suppose that a triangular truss, Fig. 69, is loaded with*

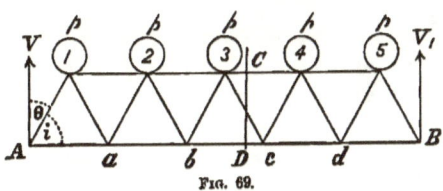

Fig. 69.

equal weights at the upper apices; it is required to find the stress upon any of the pieces of the truss.

[The *stress* is the pull or push on a piece.]

Let the truss be supported at its ends, and let

$l = Aa = ab =$ etc., $=$ the equal divisions of the span AB;
$N =$ the number of bays in the chord AB;
$L = Nl = AB$, the span;
p_1, p_2, p_3, etc., be the weights on the successive apices; which we will suppose are equal to each other; hence
$p = p_1 = p_2 =$ etc.;
$Np =$ the total load;
$V =$ the reaction at A; and
$V_1 =$ "   "   " B.

1st. *There will be equilibrium among the external forces.*

All the forces being vertical, their horizontal components will be zero, hence

$$X = 0;$$
$$Y = V + V_1 - \Sigma p = V + V_1 - Np = 0; \qquad (a)$$

and taking the origin of moments at B, observing that the moment of the load is the total load multiplied by the horizontal distance of its centre of gravity from B, we have

$$- V \cdot AB + Np \cdot \tfrac{1}{2} AB = 0;$$

or,

$$V \cdot L - Np \cdot \tfrac{1}{2} L = 0;$$
$$\therefore V = \tfrac{1}{2} Np;$$

which in (a) gives V_1 also equal to $\tfrac{1}{2}Np$; hence the supports sustain equal amounts, as they should, since the load is symetrical in reference to them, and is independent of the form of trussing.

2d. *To determine the internal forces.*—*Conceive that the truss is cut by a vertical plane and either part removed while we consider the remaining part. To the pieces in the plane section, apply forces acting in such a manner as to produce the same strains as existed before they were severed. Consider the forces thus introduced as external, and the problem is reduced to that of determining their value so that there shall be equilibrium among the new system of external forces.*

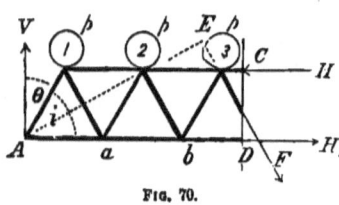

Fig. 70.

Let CD, Fig. 69, be a vertical section, and suppose that the right-hand part is removed. Introduce the external forces in place of the strains, as shown in Fig. 70.

Let H = the compressive strain in the upper chord;
H_1 = the tensile strain in the lower chord;
F = the pull in the inclined piece;
θ = the inclination of F to the vertical;
n = the number of the bay, bD, counting from A
 (which in the figure is the 3d bay); and
$D = CD$ = the depth of the frame.

The origin of coördinates may be taken at any point. Take it at A, x being horizontal and y vertical.

Resolving the forces, we have

$$X = H_1 - H + F \sin \theta = 0 ; \qquad (a)$$
$$Y = V - p_1 - p_2 - \text{etc., to } p_n - F \cos \theta = 0 ;$$
$$ = V - np - F \cos \theta = 0 ; \qquad (b)$$

the moments $= V.0 - p_1.\tfrac{1}{2}l - p_2.\tfrac{3}{2}l - \text{etc.} + H.D - F.AE = 0$,

or, $\qquad -\tfrac{1}{2}n^2pl + HD - Fnl \cos \theta = 0 \qquad (c)$

Eliminating F between equations (b) and (c), substituting the value of $V = \tfrac{1}{2}Np$, and reducing, give

$$H = \frac{pl}{2D} n(N - n); \qquad (d)$$

that is, *the strains on the bays of the upper chord vary as the product of the segments into which the lower chord is divided by the joint directly under the bay considered.*

From (b) we have

$$F \cos\theta = V - np = \tfrac{1}{2}(N - 2n)p; \qquad (e)$$

and since θ is constant, *the stress on the inclined pieces decreases uniformly from the end to the middle.*

At the middle $n = \tfrac{1}{2}N$, and $F = 0$; hence, *for a uniform load, there is no stress on the central braces.*

If F were considered as a *push*, equation (e) would be negative.

Eliminating H and F from (a), we have

$$H_1 = \left\{ N(n - \tfrac{1}{2}) - n(n - 1) \right\} \frac{pl}{2D} \qquad (f)$$

For forces in a plane the conditions of statical equilibrium give only three independent equations, (a), (b) and (c); (or Eqs. (91)); hence, *if a plane section cuts more than three independent pieces in a frame, the stresses in that section are indeterminate,* unless a relation can be established among the stresses, or a portion of them be determined by other considerations.

21. If $N = 7, p = p_1 = p_2 = $ etc. $= 1{,}000$ lbs., $AB = 56$ feet and $D = 4$ feet; required the stress on each piece of the frame.

22. In Fig. 69, if p_1 and p_2 are removed, and $p_3 = p_4 = p_5 = 1{,}000$ lbs., find the stress on the bay $2 - 3$, and the tie $2 - b$.

23. If all the joints of the lower chord are equally loaded, and no load is on the upper chord, required the stress on the nth pair of braces, counting from A, Fig. 69.

Ans. $\tfrac{1}{2}(N - 2n + 1)p \sec\theta$.

24. *A roof truss ADB is loaded with equal weights at the equidistant joints* 1, 2, 3, *etc. ; required the stress on any of its members.*

[OBS. A load composed of equal weights on all the joints will produce the same stress as that of a load uniformly distributed, except that the latter would produce cross strains upon the rafters, which it is not our purpose to discuss in this place.]

Let the tie AB be divided into equal parts, Aa, ab, etc., and the joints connected as shown in the figure. The joints are

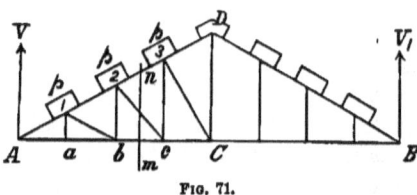

FIG. 71.

assumed to be perfectly flexible. The right half of Fig. 71 may be trussed in any manner by means of ties or braces, or both, and yet not affect the analysis applied to the left half.

Conceive a vertical section nm and the right-hand part removed. Introduce the forces H, H_1 and F as previously explained, and the conditions of the problem will be represented by Fig. 72. The letters of reference given below involve both figures.

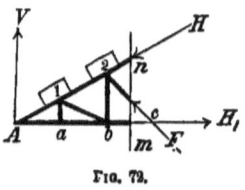

FIG. 72.

Let $N =$ the number of equal divisions (bays) in AB;
$n =$ the number of the bay bc counting from A;
$l = Aa = ab$, etc.;
$p =$ the weight on any one of the joints of the rafter;
$V =$ the vertical reaction at A or B;
$D = DC$, the depth at the vertex;
$\theta = b2c$; and $i = DAC$.

Then

$(N-1)p =$ the total load ; $\therefore V = \tfrac{1}{2}(N-1)p.$

Take the origin of coördinates at A, and the *origin of moments* at the joint marked 2. Resolving the forces shown in Fig. 72 horizontally and vertically, we have

$$X = H\cos(180° + aA1) + H_1\cos 0° + F\sin(-b2c);$$
$$Y = V - (n-1)p + H\sin(180° + aA1) + H_1\sin 0° + F\cos(-b2c);$$

or,

$$-H\cos i + H_1 - F\sin \theta = 0;$$
$$V - (n-1)p - H\sin i + F\cos \theta = 0;$$

also the *moments*,

$$H_1 b2 - V.Ab + (n-1)p.\tfrac{1}{2}(n-2)l = 0.$$

But from Fig. 71 we have

$$\frac{b2}{CD} = \frac{Ab}{AC} = \frac{(n-1)l}{\tfrac{1}{2}Nl}.$$

Substituting in the equation of moments the value of $b2$ found above, of $V = \tfrac{1}{2}(N-1)p$, of $Ab = (n-1)l$, and reducing, give

$$H_1 = \frac{Nl}{4D}(N - n + 1)p.$$

By means of the other two equations, we find

$$H = \tfrac{1}{2}(N-n)p \operatorname{cosec} i;$$
$$F = \tfrac{1}{2}(n-1)p \sec \theta.$$

e. STRESS IN A LOADED BEAM.

25. *Suppose that a beam is firmly fixed in a wall at one end, and that the projecting end is loaded with a weight P; required the forces in a vertical section mn, Fig. 73.*

Take the origin of coördinates at A, x horizontal and y vertical. Take the plane section perpendicular to the axis of x. Without assuming to know the directions in which the

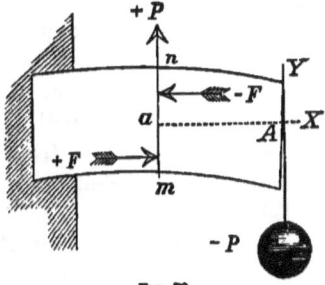

FIG. 73.

forces in the section act, we may conceive them to be resolved into horizontal and vertical components. Let F be the *typical* horizontal force, then will

$$X = \Sigma F = 0;$$

hence, some of the F-*forces* will be positive, and the others negative.

Neglecting the weight of the beam, and letting Y_1 be the sum of the vertical components in nm, we have

$$Y = Y_1 - P = 0 \quad \therefore \quad Y_1 = P;$$

as shown in the figure.

The forces, $+ P$ and $- P$, constitute a *couple* whose arm is Aa; and since the F-*forces* are the only remaining ones, the resultant of the $+ F$'*s* and the $- F$'*s* must constitute a couple whose moment equals $P.Aa$ with a contrary sign.

[OBS. Investigations in regard to the distribution of the forces over the plane section belong to the *Resistance of Materials*.]

f. LOADED CORD.

26. *Suppose that a perfectly flexible, inextensible cord is fixed at two points and loaded continuously, according to any law; it is required to find the equation of the curve and the tension of the cord.*

Assuming that equilibrium has become established, we may treat the problem as if the cord were rigid, by considering the curve which it assumes as the locus of the point of application of the resultant. The

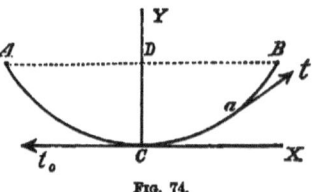

Fig. 74.

resultant at any point will be in the direction of a tangent to the curve at that point; for otherwise it would have a normal component which would tend to change the form of the curve.

Take the origin of coördinates at the lowest point of the curve. Let a be *any* point whose coördinates are x and y;

$X =$ the sum of the x-*components* of all the external forces between the origin and a;

$Y =$ the sum of the y-*components*;

$t =$ the tension of the cord at a;
$t_0 =$ the tension at the origin.

Resolving the tension (t) by multiplying it by the direction-cosine, we have

$$t \frac{dx}{ds} = \text{the } x\text{-component of } t, \text{ and}$$

$$t \frac{dy}{ds} = \text{the } y\text{-component.}$$

For the part Ca, equations (91) become

$$\left. \begin{aligned} -t_0 + X + t \frac{dx}{ds} &= 0; \\ Y + t \frac{dy}{ds} &= 0; \\ Xy - Yx &= 0. \end{aligned} \right\} \quad (a)$$

[OBS. In the problems which we shall consider, the third of these equations will be unnecessary, since the other two furnish all the conditions necessary for solving them.]

Let all the applied forces be vertical.

Then $X = 0$, and the first two of equations (a) become

$$\left. \begin{aligned} -t_0 + t \frac{dx}{ds} &= 0; \\ Y + t \frac{dy}{ds} &= 0. \end{aligned} \right\} \quad (b)$$

From the first of these we have

$$t \frac{dx}{ds} = t_0 = a \text{ constant};$$

hence, the *horizontal component of the tension will be constant throughout the length for any law of vertical loading.*

From the second of (b), we have

$$t \frac{dy}{ds} = -Y;$$

hence, the *vertical component of the tension at any point equals the total load between the lowest point and the point considered.*

27. *Let the load be uniformly distributed over the horizontal.*

(This is approximately the condition of the ordinary suspension bridge.)

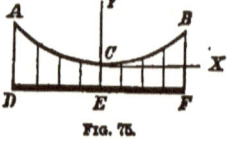

Fig. 75.

Let w = the load per unit of length, then

$$Y = -wx;$$

and (*b*) becomes

$$\left.\begin{array}{l}-t_0 + t\dfrac{dx}{ds} = 0;\\[4pt]-wx + t\dfrac{dy}{ds} = 0.\end{array}\right\} \qquad (c)$$

Eliminating t gives

$$t_0 dy = wx\,dx;$$

and integrating gives

$$t_0 y = \tfrac{1}{2}wx^2 + (C = 0);$$

$$\therefore\ x^2 = \frac{2t_0}{w} y; \qquad (d)$$

hence, the curve is a parabola whose axis is vertical, and whose parameter is $\dfrac{2t_0}{w}$. The parameter will be constant when $t_0 \div w$ is constant; hence *the tension at the lowest point will be the same for all parabolas having the same parameter and the same load per unit along the horizontal, and is independent of the length of the curve.*

To find the tension at the lowest point, substitute in equation (*d*) the value of the coördinates of some known point. Let the coördinates of the point A be x_1 and y_1; then (*d*) gives

$$t_0 = \frac{wx_1^2}{2y_1}. \qquad (e)$$

[86.] THE CATENARY. 133

To find the tension at any point we have from the first of equations (c) and the Theory of Curves

$$t = t_0 \frac{ds}{dx} = t_0 \frac{\sqrt{dx^2 + dy^2}}{dx} = t_0 \sqrt{1 + \frac{dy^2}{dx^2}}$$

$$= \frac{wx_1^2}{2y_1}\sqrt{1 + \frac{dy^2}{dx^2}}. \qquad (f)$$

To find the tension at the highest point A, from (d) find

$$\frac{dy}{dx} = \frac{2y_1}{x_1}; \qquad (g)$$

substitute in (f), and we obtain

$$t_1 = \frac{wx_1}{2y_1}\sqrt{x_1^2 + 4y_1^2}.$$

(To find t_0 by the Theory of Moments, take the origin at A. The load on x_1 will be wx_1, and its arm the horizontal distance to the centre of gravity of the load, or $\tfrac{1}{2}x_1$; hence, its moment will be $\tfrac{1}{2}wx_1^2$. The moment of the tension will be $t_0 y_1$; hence,

$$t_0 y_1 = \tfrac{1}{2}wx_1^2 \text{ or } t_0 = \frac{wx_1^2}{2y_1}, \text{ as before.})$$

The slope (or inclination of the curve to the horizontal) may be found from equation (g); which gives

$$\tan i = \frac{2y_1}{x_1}.$$

28. THE CATENARY. A catenary is the curve assumed by a perfectly flexible string of uniform section and density, when suspended at two points not in the same vertical. Mechanically speaking the load is uniformly distributed over the arc, and hence varies directly as the arc.

To find the equation,
let $w =$ the weight of the cord per unit of length;
∴ $Y = -ws$ (s being the length of the arc);
and equations (b) become

$$\left.\begin{array}{l} -t_0 + t\dfrac{dx}{ds} = 0; \\[4pt] -ws + t\dfrac{dy}{ds} = 0. \end{array}\right\} \qquad (h)$$

Transposing and dividing the second by the first, gives

$$\frac{dy}{dx} = \frac{w}{t_0} s;$$

and differentiating, substituting the value of ds and reducing, give

$$d\left(\frac{dy}{dx}\right) = \frac{w}{t_0} ds = \frac{w}{t_0} dx \sqrt{1 + \frac{dy^2}{dx^2}};$$

$$\therefore \frac{w}{t_0} dx = \frac{d\left(\frac{dy}{dx}\right)}{\sqrt{1 + \frac{dy^2}{dx^2}}}.$$

Integrating gives

$$\frac{w}{t_0} x = \log_e \left[\frac{dy}{dx} + \sqrt{1 + \frac{dy^2}{dx^2}}\right];$$

or, passing to exponentials, gives

$$e^{\frac{w}{t_0} x} = \frac{dy}{dx} + \sqrt{1 + \frac{dy^2}{dx^2}} = \frac{dy}{dx} + \frac{ds}{dx}; \qquad (i)$$

or,

$$1 + \frac{dy^2}{dx^2} = \left[e^{\frac{w}{t_0} x} - \frac{dy}{dx}\right]^2;$$

from which we find

$$\frac{dy}{dx} = \tfrac{1}{2}\left[e^{\frac{w}{t_0} x} - e^{-\frac{w}{t_0} x}\right]; \qquad (j)$$

which integrated gives

$$y = \tfrac{1}{2}\frac{t_0}{w}\left[e^{\frac{w}{t_0} x} + e^{-\frac{w}{t_0} x}\right] + \left(C = -\frac{t_0}{w}\right) \qquad (k)$$

$$= \tfrac{1}{2}\frac{t_0}{w}\left[e^{\frac{w}{2t_0} x} - e^{-\frac{w}{2t_0} x}\right]^2; \qquad (l)$$

which is the equation of the Catenary.

Eliminating $\frac{dy}{dx}$ between equations (i) and (j), we find

$$\frac{ds}{dx} = \tfrac{1}{2}\left[e^{\frac{w}{t_0} x} + e^{-\frac{w}{t_0} x}\right];$$

the integral of which is

$$s = \tfrac{1}{2}\frac{t_0}{w}\left[e^{\frac{w}{t_0}x} - e^{-\frac{w}{t_0}x}\right] + (C = 0);\quad (m)$$

which gives the length of the curve.

The following equations may also be found

$$x = \frac{t_0}{w}\log_e\left\{\frac{ws}{t_0} + \sqrt{1 + \frac{w^2s^2}{t_0^2}}\right\};$$

$$t = \sqrt{t_0^2 + w^2s^2};$$

$$s = \sqrt{\frac{yt_0}{2w}}$$

If $\theta =$ the inclination of the curve to the vertical, then

$$x = s\tan\theta\log_e\cot\tfrac{1}{2}\theta.$$

The tensions, t and t_0, are so involved that they can be determined only by a series of approximations. The full development of these equations for practical purposes belongs to Applied Mechanics.

The catenary possesses many interesting geometrical and mechanical properties, among which we mention the following:—

The centre of gravity of the catenary is lower than for any other curve of the same length joining two fixed points.

If a common parabola be rolled along a straight line, the locus of the focus will be a catenary.

According to Eq. (k) it appears that if the origin of coördinates be taken directly below the vertex at a distance equal to $t_0 \div w$, the constant of integration will be zero. (This distance equals such a length of the cord forming the catenary as that its weight will equal the tension at the lowest point of the curve). A horizontal line through this point is the *directrix of the catenary*.

The radius of curvature at any point of the catenary equals the normal at that point, limited by the directrix.

The tension at any point equals the weight of the cord forming the catenary whose length equals the ordinate of the point from the directrix.

If an indefinite number of strings (without weight) be suspended from a catenary and terminated by a horizontal line, and the catenary be then drawn out to a straight line, the lower ends of the vertical lines will be in the arc of a parabola.

If the weight of the cord varies continuously according to any known law the curve is called *Catenarian*.

29. *To determine the equation of the Catenarian curve of uniform density in which the section varies directly as the tension.*

Let k = the variable section;
δ = the weight of a unit of volume of the cord;
c = the ratio of the section to the tension;
then

$$Y = -\int \delta k ds; \quad k = ct; \quad \therefore Y = -\delta c \int t ds;$$

which substituted in (*b*) and reduced, gives

$$\delta c y = \log_e \sec c\delta x,$$

for the required equation.

g. LAW OF LOADING.

30. *It is required to find the* LAW OF LOADING *so that the action-line of the resultant of the forces at any point shall be tangent to a given curve.*

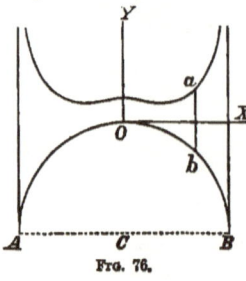

Fig. 76.

Assume the loading to be of uniform density, and the variations in the loading to be due to a variable depth. In Fig. 76, let O be the origin of coördinates; $Z = ab$ = the depth of loading over a point whose abscissa is x; d = the depth of the loading over the origin, and δ = the weight per unit of volume of the loading, then

$$Y = -\int \delta Z dx;$$

which in Eq. (*b*) gives

$$-t_0 + t \frac{dx}{ds} = 0;$$

$$-\delta \int Z\, dx + t \frac{dy}{ds} = 0;$$

Transposing, and dividing the latter by the former, gives

$$\frac{dy}{dx} = \frac{\delta}{t_0} \int Z dx;$$

which, differentiated, gives

$$\frac{d^2y}{dx^2} = \frac{\delta Z}{t_0}.$$

But, from the Theory of Curves, we have

$$\frac{d^2y}{dx^2} = \frac{\left(1 + \frac{dy^2}{dx^2}\right)^{\frac{3}{2}}}{\rho} = \frac{\sec^3 i}{\rho},$$

in which ρ is the radius of curvature, and i is the angle between a tangent to the curve and the axis of x. From these we readily find

$$Z = \frac{t_0}{\delta} \frac{\sec^3 i}{\rho}.$$

At the origin $\rho = \rho_0$, $i = 0$, and $Z = d$; which values substituted in the preceding equation give

$$\frac{t_0}{\delta} = d\rho_0;$$

$$\therefore Z = d\rho_0 \frac{\sec^3 i}{\rho}. \qquad (n)$$

Discussion. For all curves which have a vertical tangent, we have at those points

$$i = 90°; \quad \therefore \sec i = \infty, \text{ and, if } \rho \text{ is finite}$$

$$Z = \infty;$$

hence, it is practically impossible to load such a curve throughout its entire length in such a manner that the resultant shall be in the direction of the tangent to the curve. A portion of the curve, however, may be made to fulfil the required condition.

Let the given curve be the arc of a circle; then $\rho = \rho_0$, and equation (n) becomes

$$Z = d \sec^3 i,$$

from which the upper limit of the loading may be found. For

small angles $\sec^3 i$ will not greatly exceed unity, and hence, the upper limit of the load will be nearly parallel to the arc of the circle for a short distance each side of the highest point. At the extremities of the semicircle, $i = 90°$, and $Z = \infty$.

If the given curve be a parabola, we find $Z = d$, that is, the depth of loading will be constant; or, in other words, uniformly distributed over the horizontal. This is the reverse of Prob. 27.

(The principles of this topic may be used in the construction and loading of arches.)

31. *Let the tension of the cord be uniform.*

We observe in this case that the loading must act normally to the curve at every point, for if it were inclined to it, the tangential component would increase or decrease the tension.

Let $p =$ the *normal* pressure per unit of length of the arc; then $pds =$ the pressure on an *element* of length, and this multiplied by the direction-cosine which it makes with the axis of x, and the expression integrated, give

$$\int p\,ds \left(\frac{dx}{ds}\right) = \int p\,dx = \text{the } x\text{-component, and}$$

$$\int p\,dy = \text{the } y\text{-component of the pressures.}$$

hence, equations (*a*), p. 131, become

$$-t_0 + \int p\,dx + t\frac{dx}{ds} = 0;$$

$$\int p\,dy + t\frac{dy}{ds} = 0;$$

differentiating which, give

$$p\,dx + t\,d\left(\frac{dx}{ds}\right) = 0;$$

$$p\,dy + t\,d\left(\frac{dy}{ds}\right) = 0.$$

Transposing, squaring, adding and extracting the square root, give

$$p = t \left\{ \left(\frac{d^2x}{ds^2}\right)^2 + \left(\frac{d^2y}{ds^2}\right)^2 \right\}^{\frac{1}{2}} = \frac{t}{\rho}; \qquad (o)$$

that is, *the normal pressure varies inversely as the radius of curvature.*

1. If a string be stretched upon a perfectly smooth curved surface by pulling upon its two ends the normal pressure upon the surface will vary inversely as the radius of curvature of the surface, the curvature being taken in the plane of the string at that point.

2. If ρ be constant p will be constant; hence, if a circular cylinder be immersed in a fluid, its axis being vertical, the normal pressure on a horizontal arc being uniform throughout its circumference, the compression in the arc will also be constant.

h. THE LAW OF LOADING ON A NORMALLY PRESSED ARC BEING GIVEN, REQUIRED THE EQUATION OF THE ARC

32. *The ties of a suspension bridge being normal to the curve of the cable, and the load uniform along the span, required the equation of the curve of the cable.*

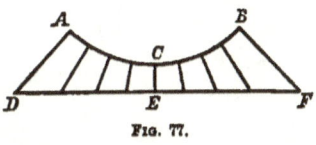

Fig. 77.

$$Ans. \left\{ 1 + \left(\frac{x}{2y} \pm \sqrt{\frac{x}{2y}}\right)^2 - 2 \right\}^{\frac{1}{2}} = \left\{ \frac{x}{2y} \pm \sqrt{\left(\frac{x}{2y}\right)^2 - 2} \right\} \cdot \frac{\rho_0}{2y};$$

the origin being at C, x horizontal and y vertical.

If $\tan i = \dfrac{dy}{dx}$, and $\rho_0 =$ the radius of curvature at the vertex, then

$$x = \tfrac{1}{2}\rho_0 (1 + \cos^2 i) \sin i,$$
$$y = \tfrac{1}{2}\rho_0 \sin^2 i \cos i.$$

(See solution by Prof. S. W. Robinson, *Journal of the Franklin Institute*, 1863, vol. 46, p. 145; and its application to bridges and arches, vol. 47, p. 152 and p. 361.)

33. *A perfectly flexible, inextensible trough of indefinite length is filled with a fluid, the edges of the trough being parallel and supported in a horizontal plane; required the equation of a cross section.*

The length is assumed to be indefinitely long, so as to eliminate the effect of the *end pieces*. The pressure of a fluid against a surface is always normal to the surface, and varies directly with the depth of the fluid. The actual pressure equals the weight of a prism of water whose base equals the surface pressed, and whose height equals the depth of the centre of gravity of the said surface below the surface of the fluid. The problem may therefore be stated as follows:—*Required the equation of the curve assumed by a cord fixed at two points in the same horizontal, and pressed normally by forces which vary as the vertical distance of the point of application below the said horizontal.*

Fig. 78.

Let A and B be the fixed points. Take the origin of the coördinates at D, midway between A and B, and y positive downwards. Let δ be the weight of a unit of volume; then

$p = \delta y$, which in equation (o) gives

$t = \delta y \rho$, and for the lowest point

$t = \delta D \rho_0$; in which D is the depth of the lowest point and ρ_0 the radius of curvature at that point;

$$\therefore \delta y \rho = \delta D \rho_0, \text{ or } \frac{y}{D\rho_0} = \frac{1}{\rho}.$$

But from the Theory of Curves we have

$$\frac{1}{\rho} = \left(1 + \frac{dy^2}{dx^2}\right)^{-\frac{3}{2}} \frac{d^2y}{dx^2};$$

which substituted above, and both sides multiplied by dy, may be put under the form

$$-\tfrac{1}{2}\left(1 + \left(\frac{dy}{dx}\right)^2\right)^{-\frac{3}{2}} d\left(\frac{dy^2}{dx^2}\right) = \frac{y\,dy}{D\rho_0};$$

the integral of which is

$$\left(1 + \frac{dy^2}{dx^2}\right)^{-\frac{1}{2}} = \frac{y^2}{2D\rho_0} + C.$$

But $\frac{dy}{dx} = 0$, for $y = D$; $\therefore C = \frac{2\rho_0 - D}{2\rho_0}$; which substituted and the equation reduced gives

$$(dx^2 + dy^2)^{\frac{1}{2}} = ds = \frac{2\rho_0 D\, dy}{\sqrt{[4D^2\rho_0^2 - (2D\rho_0 - D^2 + y^2)^2]}}.$$

Squaring and reducing, gives

$$dx = \frac{2\rho_0 D - D^2 + y^2}{\sqrt{[4D^2\rho_0^2 - (2D\rho_0 - D^2 + y^2)^2]}}\, dy.$$

These may be integrated by means of *Elliptic Functions*. Making $y = D \cos \phi$, and $c = \frac{D}{4\rho_0}$, they may be reduced to known forms. Using *Legendre's* notation, we have

$$s = \sqrt{\rho_0 D}\, \mathrm{F}_{(c,\, \phi)};$$

$$x = 2\sqrt{\rho_0 D}\left\{-\tfrac{1}{2}\mathrm{F}_{(c,\, \phi)} + \mathrm{E}_{(c,\, \phi)}\right\}.$$

(See Article by the Author in the *Journal of the Franklin Institute*, 1864, vol. 47, p. 289.)

CHAPTER V.

RELATION BETWEEN THE INTENSITIES OF FORCES ON DIFFERENT PLANES WHICH CUT AN ELEMENT.

87. DISTRIBUTED FORCES are those whose points of application are distributed over a surface or throughout a mass. The attraction of one mass for another is an example of the latter, some of the properties of which have been discussed in the Chapter on Parallel Forces; similarly, when one part of a body is subjected to a pull or push, the forces are transmitted through the body to some other part, and are there resisted by other forces. If the body be intersected by a plane, the forces which pass through it will be distributed over its surface. *Planes having different inclinations being passed through an element, it is proposed to find the relation between the intensities of the forces on the different planes.*

88. DEFINITIONS. *Stresses* are forces distributed over a surface. In the previous chapters we have *assumed* that forces are applied at points, but in practice they are always *distributed*.

A *strain* is the distortion of a body caused by a stress. Stresses tend to change the form or the dimensions of a body. Thus, a *pull* elongates, a *push* compresses, a *twist* produces torsion, etc. (See *Resistance of Materials*.)

A *simple stress* is a pull or thrust. Stresses may be compound, as a combination of a twist and a pull.

A DIRECT *simple* stress is a pull or thrust which is normal to the plane on which it acts.

A *pull* is considered *positive*, and a *push*, *negative*.

The *intensity of a stress* is the force on a unit of area, if it be constant; but, if it be *variable*, it is the ratio of the stress on an elementary area to the area.

To form a clear conception of the forces to which an element is subjected, conceive it to be removed from the body and then subjected to such forces as will produce the same strain that it had while in the body.

89. FORMULAS FOR THE INTENSITY OF A STRESS. Let F be a direct simple stress acting on a surface whose area is A, and p the measure of the intensity, then

$$p = \frac{F}{A}, \qquad (92)$$

when the stress is uniform, and

$$p = \frac{dF}{dA}, \text{ when it is variable.}$$

If the stress be variable we will assume that the section is so small that the stress may be considered uniform over its surface.

90. DIRECT STRESS RESOLVED. Let the prismatic element AB, Fig. 79, be cut by an oblique plane DE. Let the stress F be simple and direct on the surface CB, and

$N =$ the normal component of F on DE;

$T =$ the component of F along the plane DE, which is called the *tangential component;*

$\theta = FON =$ the angle between the action-line of the force and a normal to the plane DE, and is called the *obliquity of the plane;*

$A =$ the area of CB, and A' that of DE.

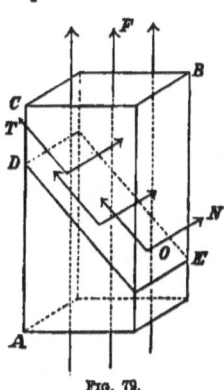

FIG. 79.

Then, according to equations (62), we have

$$N = F \cos \theta;$$
$$T = F \sin \theta.$$

From the figure we have

$$A' = A \sec \theta,$$

hence, on the plane DE, we have

$$\left.\begin{array}{l}\text{Normal intensity, } p_n = \dfrac{N}{A'} = \dfrac{F\cos\theta}{A\sec\theta} = p\cos^2\theta\,;\\[1em] \text{Tangential intensity, } p_t = \dfrac{T}{A'} = \dfrac{F\sin\theta}{A\sec\theta} = p\sin\theta\cos\theta.\end{array}\right\} \quad (93)$$

Pass another plane perpendicular to DE, having an obliquity of $90° - \theta$; then, accenting the letters, we have

$$\left.\begin{array}{l}p'_n = p\sin^2\theta\,;\\ p'_t = p\cos\theta\sin\theta.\end{array}\right\} \quad (94)$$

This result is the same as if a direct stress acting upon a plane perpendicular to CB, having an obliquity of $90° - \theta$ in reference to DE, be resolved normally and tangentially to the latter.

Combining equations (93) and (94) we readily find

$$\left.\begin{array}{l}p_n + p'_n = p\,;\\ p_t = p'_t;\end{array}\right\} \quad (95)$$

that is, *when an element (or body) under a direct simple stress is intersected by two planes the sum of whose obliquities is 90 degrees, the sum of the intensities of the normal components of the stress equals the intensity of the direct simple stress, and the intensities of the tangential stresses are equal to each other.*

91. SHEARING STRESS. The tangential stress is commonly called a *shearing stress*. It tends to draw a body sidewise along its plane of action, or along another plane parallel to its plane of action. Its action may be illustrated as follows :— Suppose that a pile composed of thin sheets or horizontal layers of paper, boards, iron, slate, or other substance, having friction between the several layers, be acted upon by a horizontal force applied at the top of the pile, tending to move it sidewise. It will tend to draw each layer upon the one immediately beneath it, and the total force exerted between each layer will equal the applied force, and the resistance at the bottom of the pile will be equal and opposite to that of the applied force. If other horizontal forces are applied at different points along the vertical face of the pile, the total tangential force at the base of the pile will equal the algebraic sum of all the applied forces.

A shearing stress and the resisting force constitute a couple,

[91.] SHEARING STRESS. 145

and as a single couple cannot exist alone, so a pair of shearing stresses necessitate another pair for equilibrium.

When the direct simple stresses on the faces of a rectangular parallelopipedon are of equal intensity, the shearing stresses will be of equal intensity.

Let Fig. 80 represent a paralellopipedon with direct and shearing stresses applied to its several faces. At present suppose that all the forces are parallel to the plane of one of the faces, as *abfe*, and call it *a plane of the forces;* then will the *planes of action*, which, in this case, will be four of the faces of the parallelopipedon, be perpendicular to *a plane of the forces.*

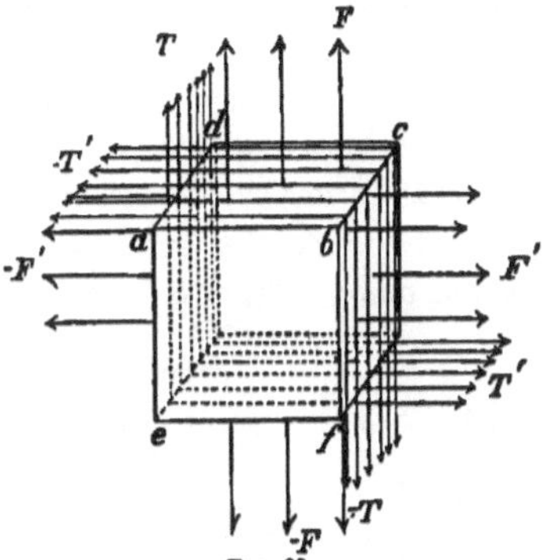

FIG. 80.

If the direct stress $+F = -F$, and $+F' = -F'$, they will equilibrate each other. The moment of the tangential force T, will be

$$p_t \times area\ fc \times ab\ ;$$

and of T'

$$p'_t \times area\ ac \times bf.$$

The couple $T.ab$ tends to turn the element to the right and $T'.bf$ to the left, hence, for equilibrium, we have

$$p_t \times area\ fc \times ab = p'_t \times area\ ac \times bf\ ;$$

but $area\ fc \times ab = area\ ac \times bf =$ the volume of the element, hence

$$p_t = p'_t. \qquad (96)$$

The effect of a pair of shearing stresses is to distort the element, changing a rectangular one into a rhomboid, as shown in Fig. 81.

Direct stresses are directly opposed to each other in the same plane or on opposite surfaces; shearing stresses act on parallel planes not coincident.

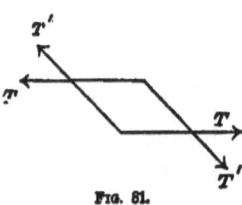

FIG. 81.

10

92. NOTATION. A very good notation was devised by Coriolis, which has been used since 1837, and is now commonly employed for the *general* investigations on this subject. It is as follows:—

Let p be a *typical* letter to denote the intensity of a stress of some kind; p_x the intensity of a stress on a plane normal to x; p_{xx} the intensity of a stress on a plane normal to x and in a direction parallel to x, and hence indicates the intensity of a *direct simple stress;* and p_{xy} the intensity of a stress on a plane normal to x but in the direction of y, and hence indicates the intensity of a *shearing stress.* Or, generally, *the first sub-letter indicates a normal to the plane of action and the second one the direction of action.* Hence we have

INTENSITIES OF THE FORCES

parallel to

$$\left.\begin{array}{ccc} x & y & z \\ p_{xx} & p_{xy} & p_{xz} \\ p_{yx} & p_{yy} & p_{yz} \\ p_{zx} & p_{zy} & p_{zz} \end{array}\right\} \text{on a plane normal to} \left\{\begin{array}{l} x\,; \\ y\,; \\ z. \end{array}\right.$$

If direct stresses only are considered, one sub-letter is sufficient; as $p_x, p_y,$ or p_z.

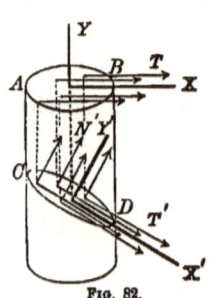

Fig. 82.

93. TANGENTIAL STRESS RESOLVED. Let T be the tangential stress on the right section $AB = A$, the section being normal to y, then

$$p_{yx} = T \div A.$$

Let CD be an oblique section, normal to the axis y'; x' and x being in the plane of the axes y and y'; then will the angle between y and y' be the obliquity of the plane CD. This we will denote by (yy'). Let the tangential force be parallel to the axis of x. Resolving this force, we have

Normal component on $CD = T \sin (yy')$;
Tangential component on $CD = T \cos (yy')$.

Dividing each of these by area $CD = AB \div \cos(yy')$, we have

$$\text{Normal intensity} = p_{y'y'} = \frac{T\sin(yy')\cos(yy')}{\text{area } AB} = p_{yx}\sin(yy')\cos(yy');$$
$$\text{Tangential intensity} = p_{y'x'} = \frac{T\cos^2(yy')}{\text{area } AB} = p_{yx}\cos^2(yy');$$
(97)

and for a tangential stress on a plane normal to x, resolved upon the same oblique plane CD, we have

$$p'_{y'y'} = p_{xy}\cos(yy')\sin(yy');$$
$$p'_{y'x'} = p_{xy}\sin^2(yy').$$
(98)

If the tangential stresses on both planes (one normal to y, and the other normal to x) are alike, and the obliquity of the plane CD less than 90°, the resultant of their tangential components will be the difference of the two components, as given by equations (97) and (98); that is, it will be $p_{y'x'} - p'_{y'x'}$; but the normal intensity will be the sum of the components as given by the same equations. The reverse will be true in regard to the *direct stresses*.

94. *Let a body be subjected to a direct simple stress; it is required to find the stresses on any two planes perpendicular to one another and to the plane of the forces; also the intensity of the stress on a third plane perpendicular to the plane of the forces; and the normal and tangential components on that plane.*

Let the forces be parallel to the plane of the paper; AO and OB, planes perpendicular to one another and to the plane of the paper, having any obliquity with the forces. Let the axis of x coincide with OB, and y with AO. Let AB be a third plane, also perpendicular to the plane of the paper, cutting the other planes at any angle. Take y' perpendicular to AB and x' parallel to it and to the plane of the paper.

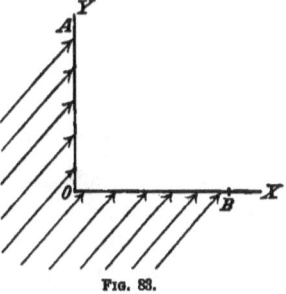

Fig. 83.

The oblique forces may be resolved normally and tangentially to the planes AO and OB, by means of equations (93) and (94). The problem will then be changed to that shown in Fig. 84, in which one set of stresses is simple and direct, and the other set tangential; and, according to Article 91, the *intensity* of the shearing stress on the two planes will be the same; hence, for this case

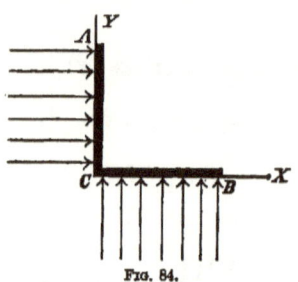

Fig. 84.

$$p_{xy} = p_{yx}$$

The total normal stress on the plane AB will be the sum of the normal components given by equations (93), (94), (97) and (98), and the total tangential stress will be the sum of the components of the tangential stress given by the same equations; hence

$$\left. \begin{array}{l} p_{y'y'} = p_{xx} \cos^2(yy') + p_{yy} \sin^2(yy') + 2p_{xy} \sin(yy') \cos(yy') ; \\ p_{y'x'} = \{p_{yy} - p_{xx}\} \sin(yy') \cos(yy') + p_{xy}\{\cos^2(yy') - \sin^2(yy')\}. \end{array} \right\} \quad (99)$$

The *resultant* stress on AB will be, according to equation (46), θ being 90°,

$$p_r = \sqrt{p^2_{y'y'} + p^2_{y'x'}} ; \qquad (100)$$

and the inclination of the resultant stress to the normal, y', will be

$$\tan(ry') = \frac{p_{y'x'}}{p_{y'y'}}. \qquad (101)$$

95.   Discussion of equations (99).

A. Find the inclination of the plane on which there is no tangential stress.

In the 2d of equations (99) make $p_{y'x'} = 0$, and representing this particular angle by (YY'), we find

$$\tan 2(\text{YY}') = \frac{2 \sin(yy') \cos(yy')}{\cos^2(yy') - \sin^2(yy')} = \frac{2p_{xy}}{p_{xx} - p_{yy}}, \quad (102)$$

which gives two angles differing from each other by 90°, or, the planes will be perpendicular to one another.

Hence, *in every case of a direct simple stress upon a pair of planes perpendicular to one another and to a plane of the stresses, there are two planes, also perpendicular to one another and to the plane of the stresses, on which there is no tangential stress.*

These two directions are called *principal axes of stress.*

Principal axes of stress are the normals to two planes perpendicular to one another on which there is no tangential stress.

Principal stresses are such as are parallel to the principal axes of stress. (In some cases there is a third principal stress perpendicular to the plane of the other two.)

The formulas for the stresses become most simple by referring them directly to the principal axes.

a. Let one of the direct stresses be zero, or $p_{YY} = 0$.
Equation 102 gives

$$\tan 2(YY') = \frac{2p_{xy}}{p_{xx}} \qquad (103)$$

b. Let one of the direct stresses be a pull, and the other a push.

Then p_{YY} becomes negative, and we have

$$\tan 2(YY') = \frac{2p_{xy}}{p_{xx} + p_{YY}} \qquad (104)$$

c. Let them act in opposite senses and equal to each other.
Then

$$\tan 2(YY') = \frac{p_{xy}}{p_{xx}}. \qquad (105)$$

d. Let there be no tangential stress on the original planes, or $p_{xy} = 0$.
Then,
$$\tan 2(YY') = 0; \quad \therefore (YY') = 0 \text{ or } 90°;$$
and the original planes are *principal planes.*

e. *Let there be no direct stresses.*

Then,

$$\tan 2(\text{YY}') = \infty\,;\ \text{or}\ (\text{YY}') = 45° \text{ or } 135°\,;\quad (106)$$

that is, *if on two planes, perpendicular to one another and to the plane of the stresses, there are no direct stresses, then will the stress on two planes, perpendicular to one another and to the plane of the stresses, whose inclination with the original planes is* 45°, *be simple and direct.*

f. Let the direct stresses be equal to one another and act in the same sense, and let there be no shearing on the original planes.

Then

$$\tan 2(\text{YY}') = \frac{0}{0}\,;$$

and (YY′) is indeterminate; hence, in this case every plane perpendicular to a plane of the stress will be a principal plane.

EXAMPLES.

1. A rough cube, whose weight is 550 pounds, rests on a horizontal plane. A stress of 150 pounds applied at the upper face pulls vertically upward, and another direct stress of 125 pounds, applied at one of the lateral faces, tends to draw it to the right, while another direct stress of 50 pounds tends to draw it to the left; required the position of the planes on which there are no tangential stresses.

If the cube is of finite size it will be necessary to modify the problem, in order to make it agree with the hypothesis under which the formulas have been established. The force of gravity being distributed throughout the mass, would cause a variable stress, and the surface of no shear would be curved instead of plane. We will therefore assume *that the cube is without weight, and the* 550 *pounds is applied directly to the lower surface.* Then the vertical stress will be 150 pounds, the remaining 400 pounds being resisted directly by the plane on which it rests, and so far as the present problem is concerned, only produces friction for resisting the shearing stress. The direct horizontal stress will be 50 pounds, the remaining 75 pounds producing a shearing on the horizontal plane. The former force tends to turn the cube right-handed by rotating it about the lower right-hand corner, thus producing a reaction or vertical tangential stress of 75 pounds. Let the area of each face of the cube be unity, then we have

$p_{xy} = 75$ pounds; $p_{xx} = 50$ pounds; $p_{yy} = 150$ pounds;

and these in (102) give

$$\tan 2(YY') = \frac{2 \times 75}{50 - 150} = -1.5;$$

$$\therefore (YY') = -28°\ 9'\ 18'', \text{ or } +61°\ 50'\ 42''.$$

If the body be divided along either of these planes, the forces will tend to lift one part directly from the other without producing sliding upon the plane of division.

2. A rough body, whose weight is 100 pounds, rests on an inclined plane; required the normal and tangential components on the plane. (Use Eq. (93).)

3 A block without weight is *secured* to a horizontal plane and thrust downward by a stress whose intensity is 150 pounds, and pulled towards the right by a stress whose intensity is 150 pounds, and to the left with an intensity of 100 pounds; required the plane of no shear.

4. A cube rests on a horizontal plane, and one of its vertical faces is forced against a vertical plane by a stress of 200 pounds applied at the opposite face, and on one of the other vertical faces is a direct pulling stress of 75 pounds, which is directly opposed by a stress of 50 pounds on the opposite vertical face; required the position of the plane of no shear.

In this case the weight of the cube would be a third principal stress, but it is eliminated by the conditions of the problem. The shearing stress is 25 pounds; and because the direct stresses are unlike, we use Eq. (104).

5. A rectangular parallelopipedon stands on a horizontal plane, and on the opposite pairs of vertical faces tangential

stresses of equal intensities are applied; required the position of the plane of no shear. (See Eq. (106).)

6. In the preceding problem find the intensity of the direct stress on the plane of no shear. (Substitute the proper quantities in the 1st of (99).)

B. To find the planes of action for maximum and minimum normal stresses, and the values of the stresses.

Equate to zero the first differential coefficient of the 1st of Equations (99), and we have

$$\left. \begin{array}{l} -2p_{xx}\cos(yy')\sin(yy') + 2p_{yy}\sin(yy')\cos(yy') \\ -2p_{xy}\sin^2(yy') + 2p_{xy}\cos^2(yy') = 0 \end{array} \right\} \quad (107)$$

$$\therefore \tan 2(yy') = \frac{2p_{xy}}{p_{xx} - p_{yy}};$$

which, being the same as (102), shows that on those planes which have no shearing stress, the direct stress will be either a maximum or a minimum. Testing this value by the second differential coefficient, we find that one of the values of (YY') gives a maximum and the other a minimum.

Comparing (107) with the second of (99), shows that *the first differential coefficient of the value of the direct stress on any plane equals the shearing stress on that plane.*

From (107), observing that $\cos(yy') = \sqrt{1 - \sin^2(yy')}$, we find

$$\sin^2(YY') = \frac{1}{2} \left\{ 1 \pm \frac{p_{yy} - p_{xx}}{\sqrt{(p_{yy} - p_{xx})^2 + 4p^2_{xy}}} \right\}; \quad (108)$$

and these values in the 1st of (99), and the maximum and minimum values designated by $p_{Y'}$, give

$$p_{Y'} = \tfrac{1}{2}(p_{xx} + p_{yy}) \pm \sqrt{\tfrac{1}{4}(p_{xx} - p_{yy})^2 + p^2_{xy}}; \quad (109)$$

in which the upper sign gives the *maximum*, and the lower the *minimum* stress. These are *principal stresses*, and we denote them by one sub-letter.

MAXIMUM STRESS.

a. If $p_{xy} = 0$, we have

$$(yy') = 0° \text{ or } 90°, \text{ as we should.}$$

b. If $p_{yy} = 0$, we have

$$\left.\begin{array}{l} maximum, p_{Y'} = \tfrac{1}{2}p_{xx} + \sqrt{\tfrac{1}{4}p^2_{xx} + p^2_{xy}}\,; \\ minimum, p_{Y'} = \tfrac{1}{2}p_{xx} - \sqrt{\tfrac{1}{4}p^2_{xx} + p^2_{xy}}\,; \end{array}\right\} \quad (110)$$

hence, the maximum normal stress will be of the same *kind* as the principal direct stress, p_{xx}; that is, if the latter is a *pull*, the former will also be a *pull*, and the minimum principal stress will be of the opposite kind.

c. If there are no direct stresses p_{xx} will also be zero, and we have

$$(YY') = 45° \text{ or } 135°;$$

and

$$maximum\ p_{Y'} = p_{xy} = -p_{Y'}\ for\ minimum\,;$$

that is, the principal stresses will have the same intensity as the shearing stresses, and act on planes perpendicular to one another, and inclined 45° to the original planes.

EXAMPLES.

1. Suppose that a rectangular box rests on one end, and that one pair of opposite vertical sides press upon the contents of the box with an intensity of 20 pounds, and the other pair of vertical faces press with an intensity of 40 pounds, and that horizontal tangential stresses, whose intensities are 10 pounds, are applied to the vertical faces, one pair tending to turn it to the right, and the other to the left; required the position of the vertical planes of no shearing, and the maximum and minimum values of the direct stresses.

2. For an application of Equations (103) and (110) to the stresses in a beam, see the Author's *Resistance of Materials*, 2d edition, pp. 236-240.

C. To find the position of the planes of maximum and minimum shearing.

Equate to zero the first differential coefficient of the second of (99) and reduce, denoting the angles sought by (YY'), and we find,

$$-\cot 2(YY') = \tan 2(yy');$$

$$\therefore 2YY' = 2(yy') + 90°;$$

or,

$$YY' = yy' + 45°;$$

that is, *the planes of maximum and minimum shear make angles of 45 degrees with the* PRINCIPAL PLANES.

D. *Let the planes be* PRINCIPAL SECTIONS.

Then the stresses will be *principal stresses*, and $p_{xy} = 0$. Using a single subscript for the direct stresses, equations (99) become

$$\left.\begin{array}{l} p_{y'} = p_x \cos^2(yy') + p_y \sin^2(yy'); \\ p_{y'x'} = (p_x - p_y) \sin(yy') \cos(yy'). \end{array}\right\} \quad (111)$$

a. Let $p_x = p_y$, then

$$p_{y'} = p_x; \text{ and } p_{y'x'} = 0;$$

that is, *when two principal stresses are alike and equal on a pair of planes perpendicular to the plane of the stresses, the normal intensity on every plane perpendicular to the plane of the stresses will be equal to that on the principal planes, and there will be no shearing on any plane.*

This condition is realized in a perfect fluid, and hence very nearly so in gases and liquids, since they offer only a very slight resistance to a tangential stress. If a vessel of any liquid be intersected by two vertical planes perpendicular to one another, the pressure per square inch will be the same on both, and will be normal to the planes; hence, according to the above, it will be the same upon all planes traversing the same point. This is only another way of stating the fact that fluids press equally in all directions.

b. *To find the planes on which there will be no normal pressure.*

For this $p_{y'}$ in (111) will be zero;

$$\therefore \tan(yy') = \sqrt{\frac{p_x}{p_y}}\sqrt{-1};$$

which, being imaginary, shows that it is impossible when the stresses are alike; but if they are *unlike*, we have

$$\tan(yy') = \sqrt{\frac{-p_x}{p_y}}\sqrt{-1} = \sqrt{\frac{p_x}{p_y}}.$$

If $p_x = -p_y$, then $(yy') = 45°$, and the 2d of (111) gives

$$p_{y'x'} = p_x;$$

which shows that when the direct stresses are *unlike* and of *equal intensity* on planes perpendicular to one another, the shearing stress on a plane cutting both the others at an angle of 45 degrees, will be of the same intensity.

Let $(yy') = 45°$, or $135°$, then (111) become

$$\left. \begin{array}{l} p_{y'} = \tfrac{1}{2}(p_x + p_y); \\ p_{y'x'} = \pm \tfrac{1}{2}(p_x - p_y); \end{array} \right\} \quad (112)$$

in the latter of which the upper sign gives a maximum, and the lower a minimum value.

· Using the upper sign, we find

$$\left. \begin{array}{l} p_x = p_{y'} + p_{y'x'}; \\ p_y = p_{y'} - p_{y'x'}. \end{array} \right\} \quad (113)$$

96. Problem. *Find the plane on which the obliquity of the stress is greatest, the intensity of that stress, and the angle of its obliquity.*

Let the stresses be *principal stresses* and of the *same kind*, and ϕ the angle of obliquity of the required plane to the stress; then

$\sin \phi = \dfrac{p_x - p_y}{p_x + p_y}$; *the intensity* $= \sqrt{(p_x p_y)}$; and the angle between the principal plane x and the required plane $= 45° - \tfrac{1}{2}\phi$.

If the principal stresses are *unlike*, then

$\sin \phi' = \dfrac{p_x + p_y}{p_x - p_y}$; *the intensity* $= \sqrt{-p_x p_y}$, and the angle between the principal plane x, and the oblique plane $= 45° - \tfrac{1}{2}\phi'$.

EXAMPLE.

If a body of sand is retained by a vertical wall and the *intensity* of the horizontal push is 25 pounds, and of the vertical pressure is 75 pounds; required the plane on which the resultant has the greatest obliquity, and the intensity of the stress on that plane.

CONJUGATE STRESSES.

97. A pair of stresses, each acting parallel to the plane of action of the other, and whose action-lines are parallel to a plane which is perpendicular to the line of intersection of the planes of action, are called *conjugate stresses*.

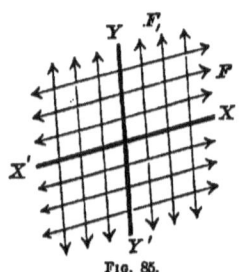

Fig. 85.

Thus, in Fig. 85, one set of stresses acts on the plane YY, parallel to the plane XX, and the other set on XX, parallel to YY. In a rigid body the intensities of these sets of stresses are independent of each other; for each set equilibrates itself. *Principal stresses* are also *conjugate*.

There may be *three conjugate stresses* in a body, and *only three*. For, in Fig. 85, there may be a third stress on the plane of the paper, which may be parallel to the line of intersection of the planes XX and YY, and each stress will be parallel to the plane of the other two. A fourth stress cannot be introduced which will be conjugate to the other three.

Conjugate stresses may be resolved into normal and tangential components on their planes of action, and treated according to the preceding articles. The fact that the stresses have the same obliquity, being the complement of the angle made by the planes, simplifies some of the more general problems of stresses.

GENERAL PROBLEM.

98. *Given the stresses on the three rectangular coördinate planes; required the stresses on any oblique plane in any required direction.*

As before, the element is supposed to be indefinitely small. Let abc be the oblique plane, the normal to which designate by n. The projection of a unit of area of this plane on each of the coördinate plains, gives respectively

$$\cos(nx),\ \cos(ny),\ \cos(nz).$$

The direct stress parallel to x acting on the area $\cos(nx)$ will give a stress of $p_{xx} \cos(nx)$, and the tangential stress normal to y and parallel to x will produce a stress $p_{yx} \cos(ny)$, and similarly the tangential stress normal to z and parallel to x gives $p_{zx} \cos(nz)$; hence the total stress on the unit normal to n and parallel to x will be

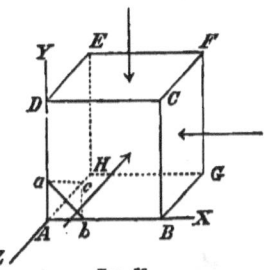

Fig. 86.

$$\left.\begin{array}{l} p_{nx} = p_{xx} \cos(nx) + p_{yx} \cos(ny) + p_{zx} \cos(nz); \\ \text{similarly,} \\ p_{ny} = p_{xy} \cos(nx) + p_{yy} \cos(ny) + p_{zy} \cos(nz); \\ p_{nz} = p_{xz} \cos(nx) + p_{yz} \cos(ny) + p_{zz} \cos(nz). \end{array}\right\} \quad (114)$$

Let these be resolved in any arbitrary direction parallel to s. To do this multiply the first of the preceding equations by $\cos(sx)$, the second by $\cos(sy)$, and the third by $\cos(sz)$, and add the results.

For the purpose of abridging the formulas, let $\cos(nx)$ be written Cnx, and similarly for the others. Then we have

$$\left.\begin{array}{l} p_{ns} = p_{xx} CnxCsx + p_{yy} CnyCsy + p_{zz} CnzCsz \\ \qquad + p_{yz}(CnyCsz + CnzCsy) + p_{xz}(CnzCsx \\ \qquad + CnxCsz) + p_{xy}(CnxCsy + CnyCsx). \end{array}\right\} \quad (115)$$

This expression being *typical*, we substitute x' for n and s, and thus obtain an expression for the intensity on a surface normal to x' and parallel to x'. Or generally, substitute successively x', y', z' for n and s, and we obtain the following formulas:

DIRECT STRESSES.

$$p_{x'x'} = p_{xx}C^2xx' + p_{yy}C^2yx' + p_{zz}C^2zx' + 2p_{yz}Cyx'Czx'$$
$$+ 2p_{zx}Czx'Cxx' + 2p_{xy}Cxx'Cyx';$$

$$p_{y'y'} = p_{xx}C^2xy' + p_{yy}C^2yy' + p_{zz}C^2zy' + 2p_{yz}Cyy'Czy'$$
$$+ 2p_{zx}Czy'Cxy' + 2p_{xy}Cxy'Cyy';$$

$$p_{z'z'} = p_{xx}C^2xz' + p_{yy}C^2yz' + p_{zz}C^2zz' + 2p_{yz}Cyz'Czz'$$
$$+ 2p_{zx}Czz'Cxz' + 2p_{xy}Cxz'Cyz';$$

TANGENTIAL STRESSES.

$$p_{y'z'} = p_{xx}Cxy'Cxz' + p_{yy}Cyy'Cyz' + p_{zz}Czy'Czz'$$
$$+ p_{yz}(Cyy'Czz' + Cyz'Czy') + p_{zx}(Czy'Cxz'$$
$$+ Czz'Cxy') + p_{xy}(Cxy'Cyz' + Cxz'Cyy');$$

$$p_{z'x'} = p_{xx}Cxz'Cxx' + p_{yy}Cyz'Cyx' + p_{zz}Czz'Czx'$$
$$+ p_{yz}(Cxz'Czx' + Cyx'Czy') + p_{zx}(Czz'Cxx' + Czx'Cxz')$$
$$+ p_{xy}(Cxz'Cyz' + Cxx'Cyz');$$

$$p_{x'y'} = p_{xx}Cxx'Cxy' + p_{yy}Cyx'Cyy' + p_{zz}Czx'Czy'$$
$$+ p_{yz}(Cyx'Czy' + Cyy'Czx') + p_{zx}(Czx'Cxy' + Czy'Cxx')$$
$$+ p_{xy}(Cxx'Cyy' + Cxy'Cyx').$$

It may be shown *that for every state of stress in a body there are three planes perpendicular to each other, on which the stress is entirely normal.*

[These equations are useful in discussing the general *Theory of the Elasticity of Bodies.*]

These formulas apply to *oblique* axes as well as *right*, only it should be observed when they are oblique that $p_{y'z'}$ is not a stress on a plane normal to y', parallel to z', but on a plane normal to x' resolved in the proper direction.

CHAPTER VI.

VIRTUAL VELOCITIES.

99. DEF. If the point of application of a force be moved in the most arbitrary manner an indefinitely small amount, the projection of the path thus described on the original action-line of the force is called a *virtual velocity*. The product of the force into the virtual velocity is called *the virtual moment*. Thus, in Fig. 87, if a be the point of application of the force F, and ab the arbitrary displacement, ac will be the virtual velocity, and $F.ac$ the virtual moment.

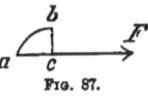

FIG. 87.

The path of the displacement must be so short that it may be considered a straight line; but in some cases its length may be finite.

If the projection falls upon the action-line, as in Fig. 87, the virtual velocity will be considered *positive*, but if on the line prolonged, it will be *negative*.

100. PROP. *If several concurring forces are in equilibrium, the algebraic sum of their virtual moments will be zero.*

Using the notation of Article (47), and in addition thereto let

l be the length of the displacement; and

p, q, and r the angles which it makes with the respective coördinate axes;

then will the projections of l on the axes be

$$l \cos p, \quad l \cos q, \quad l \cos r,$$

respectively. Multiplying equations (50), by these respectively, we have

$$F_1 \cos \alpha_1 \, l \cos p + F_2 \cos \alpha_2 \, l \cos p + \text{etc.} = 0;$$
$$F_1 \cos \beta_1 \, l \cos q + F_2 \cos \beta_2 \, l \cos q + \text{etc.} = 0;$$
$$F_1 \cos \gamma_1 \, l \cos r + F_2 \cos \gamma_2 \, l \cos r + \text{etc.} = 0.$$

Adding these together term by term, observing that

$$\cos \alpha \cos p + \cos \beta \cos q + \cos \gamma \cos r = \cos (Fl);$$

which is the cosine of the angle between the action-line of F and the line l; and that $l \cos (Fl) = \delta f$ (read, variation f) is the virtual velocity of F, we have

$$F_1 \delta f_1 + F_2 \delta f_2 + F_3 \delta f_3 + \text{etc.} = \Sigma F \delta f = 0; \quad (116)$$

which was to be proved.

101. *If any number of forces in a* SYSTEM *are in equilibrium, the sum of their virtual moments will be zero.*

Conceive that the point of application of each force is connected with all the others by rigid right lines, so that the action of all the forces will be the same as in the actual problem. If any of the lines thus introduced are not subjected to stress, they do not form an essential part of the system and may be cancelled at first, or considered as not having been introduced. Let the system receive a displacement of the most arbitary kind. At each

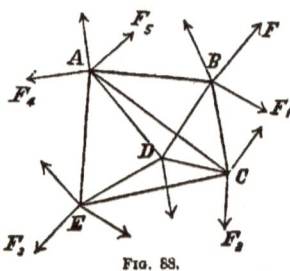

Fig. 88.

point of application of a force or forces, the stresses in the rigid lines which meet at that point, combined with the applied force or forces at the same point, are necessarily in equilibrium, and by separating it from the rest of the system, we have a system of concurrent forces. Hence, for the point B, for instance, we have, according to (116),

$$F \delta f + F_1 \delta f_1 + \text{etc.} + BC \delta BC + BA \delta BA + BD \delta BD = 0;$$

in which BC, etc., are used for the tension or compression which may exist in the line. But when the point C is considered, we will have $BC \delta BC$ with a contrary sign from that in the preceding expression, and hence their sum will be zero.

Proceeding in this way, as many equations may be established as there are points of application of the forces; and adding the equations together, observing that all the expressions which represent stresses on the lines disappear, we finally have

$$\Sigma F \delta f = 0. \quad (117)$$

The converse is evidently true, that when *the sum of the virtual moments is zero the system is in equilibrium.*

Equations (116) and (117) are no more than the *vanishing equations for work*. If a *system* of forces is in equilibrium it does no work. This principle is easily extended to Dynamics. For, the work which is stored in a moving body equals that done by the impelling force above that which it constantly does in overcoming resistances. Thus, when friction is overcome, the impelling forces accomplish work in overcoming this resistance, and all above *that* is stored in the moving mass. Letting R be the resultant of all the impressed forces producing motion, and s the path described by the body, we have

$$R\delta r - \Sigma m \frac{d^2 s}{dt^2} \delta s = 0. \qquad (118)$$

This is the most general principle of Mechanics, and M. Lagrange made it the fundamental principle of his celebrated work on *Mécanique Analytique*, which consisted chiefly of a discussion of equation (118).

EXAMPLES.

1. Determine the conditions of equilibrium of the straight lever.

Let AB be the lever, having a weight P at one end and W at the other, in equilibrium on the fulcrum G.

Conceive the lever to be turned infinitesimally about G, taking the position CD, then will Aa, which is the projection of the path AC on the action-line of P, be the virtual velocity of P; and similarly Bb will be the virtual velocity of W. The former will be positive and the latter negative; hence

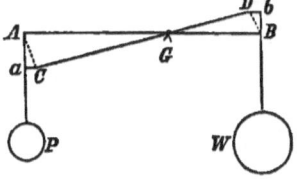

Fig. 89.

$$P.Aa - W.Bb = 0.$$

The triangles AaC and ACG at the limit are similar, having the right angles AaC and ACG, $aAC = AGC$, and the remaining angle equal. Similarly, bDB is similar to BGD.

$$\therefore \frac{Aa}{Bb} = \frac{CG}{DG} = \text{(at the limit)} \frac{AG}{BG};$$

which, substituted in the preceding expression, gives

$$P.AG = W.BG;$$

that is, *the weights are inversely proportioned to the arms.*

If the lever be turned about the end A, we would find in a similar manner that $(P + W).AG = W.AB$; in which $P + W$ is the reaction sustained by the fulcrum G.

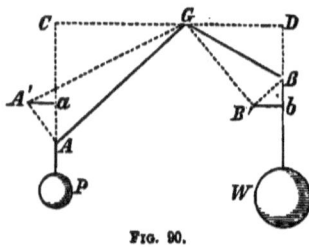

Fig. 90.

2. Find the conditions of equilibrium of the bent lever.

Let AG and GB be the arms of the lever and G the fulcrum. Let it be turned slightly about G; then will Aa and Bb be the respective virtual velocities of P and W;

$$\therefore - P.Aa + W.bB = 0.$$

From G draw GC perpendicular to PA, then will the triangle ACG be similar to AaA', having the angle $AaA' = ACG$; and $aAA' = CGA$. Similarly, the triangle BDG is similar to BbB';

$$\therefore \frac{Aa}{Bb} = \frac{GC}{GD};$$

which, combined with the preceding equation, gives

$$P.GC = W.GD;$$

that is, *the weights are inversely proportional to their horizontal distances from the fulcrum.*

3. Find the conditions of equilibrium of the single pulley.

In Fig. 91, let the weight P be moved a distance equal to ab, then will W be moved a distance $cd = ab$; hence, we have

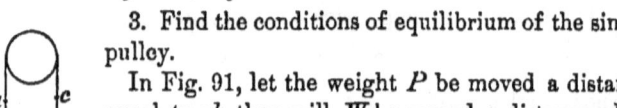

Fig. 91.

$$- P.ab + W.cd = 0; \therefore P = W.$$

4. On the inclined plane AC, a weight P is held by a force W acting parallel to the plane; required the relation between P and W.

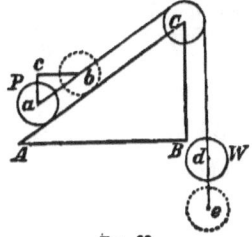

Fig. 92.

$de = ab$ will be the virtual velocity of W, and ac that of P; and we have

$$- P.ac + W.ab = 0.$$

From the similar triangles abc and ABC, we have

$$\frac{ac}{ab} = \frac{CB}{AC} \quad \therefore P.CB = W.AC; \text{ or}$$

$$P : W :: AC : CB.$$

5. On the inclined plane, if the weight P is held by a force W, acting horizontally, required the relation between P and W.

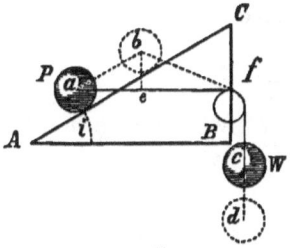

Fig. 93.

The movement being made, cd will be the virtual velocity of W, which at the limit equals ae, and be will be the virtual velocity of P, and we have

$$- P.be + W.ae = 0; \text{ and } ae : eb :: AB : BC,$$
$$\therefore P.CB = W.BA;$$

or, *the weight is to the horizontal force as the base of the triangle is to its altitude*

6. In Fig. 27 show that $Pdr = Wdy$.

7. One end of a beam rests on a horizontal plane, and the other on an inclined plane; required the horizontal pressure against the inclined plane.

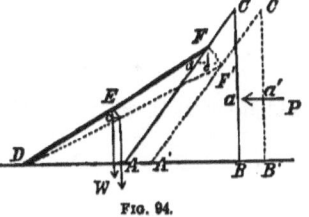

Fig. 94.

This involves the principle of the wedge; for the block ABC may represent one-half of a wedge being forced against the resistance W. Conceive the plane to be moved a distance AA', and that the beam turns

about the end D, but is prevented from sliding on the plane; then will the virtual velocity of the horizontal pressure be AA', and that of the weight will be Ec; hence, for equilibrium we have
$$W.Ec - P.AA' = 0. \qquad (a)$$
We now find the relation between Ec and AA'.

Let $l = DF$, the length of the beam;
   $a = DE$, the distance from D to the centre of gravity of the beam;
   $a = CAB$; $\quad \beta = ADF$.

The end at F will describe an arc FF' about D as a centre. From F' draw $F'd$ parallel to AA', and from F drop a perpendicular Fe upon dF'. Then, from similar triangles, we have
$$Fe = \frac{l}{a} Ec,$$
FF' will be perpendicular to DF, and Fe perpendicular to dF', hence
$$eFF' = ADF = \beta;\ dFe = 90° - a;$$
$$\therefore dFF' = 90° - a + \beta;$$
and
$$FF' = Fe \sec \beta = \frac{l}{a} Ec \sec \beta.$$

The triangle FdF' gives
$$\frac{FF'}{AA' = dF'} = \frac{\sin a}{\sin (90° - a + \beta)},$$
hence,
$$\frac{Ec}{AA'} = \frac{a}{l} \frac{\sin a \cos \beta}{\cos (a - \beta)};$$
which, substituted in equation (a) above, gives
$$P = W \frac{a}{l} \frac{\sin a \cos \beta}{\cos (a - \beta)}.$$

8. Deduce the formula for the triangle of forces from the principle of Virtual Velocities.

CHAPTER VII.

MOMENT OF INERTIA.

(This chapter may be omitted until its principles are needed hereafter (see Ch. X.) Although the expression given below, called the *Moment of Inertia*, comes directly from the solution of certain mechanical problems, yet its principles may be discussed without involving the idea of *force*, the same as any other mathematical expression. The *term* probably originated from the idea that inertia was considered *a force*, and in most mechanical problems which give rise to the *expression* the moment of a force is involved. But the expression is not in the *form* of a simple moment. If we consider a moment as the product of a quantity by an arm, it is of the *form of a moment of a moment*. Thus, dA being the quantity, ydA would be a moment, then considering this as a new quantity, multiplying it by y gives $y^2 dA$, which would be a moment of the moment. Since *we* do not consider inertia as a force, and since all these problems may be reduced to the consideration of geometrical magnitudes, it appears that some other term might be more appropriate. It being, however, universally used, a change is undesirable unless a new and better one be *universally* adopted.)

DEFINITIONS.

102. The expression, $\int y^2 dA$, in which dA represents an element of a body, and y its ordinate from an axis, occurs frequently in the analysis of a certain class of problems, and hence it has been found convenient to give it a special name. It is called *the moment of inertia*.

THE MOMENT OF INERTIA OF A BODY

is the sum of the products obtained by multiplying each element of the body by the square of its distance from an axis.

The axis is any straight line in space from which the ordinate is measured.

The quantity dA may represent an element of a line (straight or curved), a surface (plane or curved), a volume, weight, or mass; and hence the above definition answers for all these quantities.

MOMENT OF INERTIA. [103.]

The moment of inertia of a plane surface, when the axis lies in it, is called a *rectangular moment;* but when the axis is perpendicular to the surface it is called a *polar moment.*

103. EXAMPLES.

1. Find the moment of inertia of a rectangle in reference to one end as an axis.

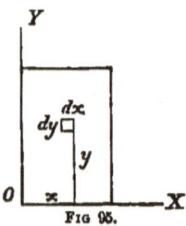
Fig 95.

Let $b =$ the breadth, and $d =$ the depth of the rectangle. Take the origin of coördinates at O.

We have $dA = dydx;$

and
$$I = \int_0^d \int_0^b y^2 dydx = b\int_0^d y^2 dy = \tfrac{1}{3}bd^3.$$

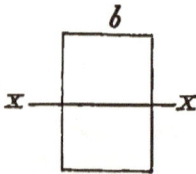
Fig. 96.

2. What is the moment of inertia of a rectangle in reference to an axis through the centre and parallel to one end?

Ans. $\tfrac{1}{12}bd^3.$

3. What is the moment of inertia of a straight line in reference to an axis through one end and perpendicular to it, the section of the line being considered unity?

Ans. $\tfrac{1}{3}l^3.$

4. Find the moment of inertia of a circle in reference to an axis through its centre and perpendicular to its surface.

We represent the *polar moment* of inertia by I_p.

Let $r =$ the radius of the circle;
$\rho =$ the radius vector;
$\theta =$ the variable angle; then
$d\rho =$ one side of an elementary rectangle;
$\rho d\theta =$ the other side; and
$dA = \rho d\rho d\theta;$

and, according to the definition, we have

$$I_p = \int_0^r \int_0^{2\pi} \rho^3 d\rho d\theta = \tfrac{1}{2}\pi r^4.$$

5. What is the moment of inertia of a circle in reference to a diameter as an axis? (See Article 105.)

Ans. $\tfrac{1}{4}\pi r^4$.

6. What is the moment of inertia of an ellipse in reference to its major axis; a being its semi-major axis and b, its semi-minor?

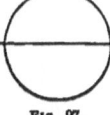

Fig. 97.

Ans. $\tfrac{1}{4}\pi ab^3$.

7. Find the moment of inertia of a triangle in reference to an axis through its vertex and parallel to its base.

Let b be the base of the triangle, d its altitude, and x any width parallel to the base at a point whose ordinate is y; then $dA = dxdy$, and we have

$$I = \int_0^d \int_0^{\tfrac{b}{d}y} y^2 dxdy = \frac{b}{d}\int_0^d y^3 dy = \tfrac{1}{4}bd^3.$$

8. What is the moment of inertia of a triangle in reference to an axis passing through its centre and parallel to the base?

Ans. $\tfrac{1}{36}bd^3$.

9. What is the moment of inertia of an isosceles triangle in reference to its axis of symmetry?

Ans. $\tfrac{1}{48}b^3 d$.

10. Find the moment of inertia of a sphere in reference to a diameter as an axis.

The equation of the sphere will be $x^2 + y^2 + z^2 = R^2$. The moment of inertia of any section perpendicular to the axis of x will be $\tfrac{1}{2}\pi y^4$; hence for the sphere we have

$$I = \int_{-R}^{+R} \tfrac{1}{2}\pi y^4 dx = \pi \int_0^R (R^2 - x^2)^2 dx = \tfrac{8}{15}\pi R^5.$$

FORMULA OF REDUCTION.

104. *The moment of inertia of a body, in reference to any axis, equals the moment of inertia in reference to a parallel axis passing through the centre of the body plus the product of the area (or volume or mass) by the square of the distance between the axes.*

This proposition for plane areas was proved in Article 80.

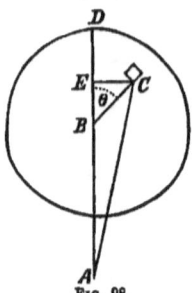
Fig. 98.

To prove it generally, let Fig. 98 represent the projection of a body upon the plane of the paper, B the projection of an axis passing through the centre of the body, A any axis parallel to it, C the projection of any element; $AC = r$, $BC = r_1$, the angle $CBD = \theta$, and $V =$ the volume of the body.

Then

$I_1 = \int r_1^2 dV$ will be the moment of inertia of the volume in reference to the axis through the centre; and

$I = \int r^2 dV$, the moment in reference to the axis through A.

Let $AB = D$, then $AE = D + r_1 \cos\theta$, and $r^2 = r_1^2 \sin^2\theta + (D + r_1 \cos\theta)^2$;

$$\therefore \int r^2 dV = \int r_1^2 dV + 2D\int r_1 \cos\theta\, dV + D^2\int dV.$$

But $\int r_1 \cos\theta\, dV = 0$, since it is the statical moment of the body in reference to a plane perpendicular to AD passing through the centre of the body and perpendicular to the plane of the paper, therefore the preceding equation becomes

$$I = I_1 + VD^2; \qquad (119)$$

which is called the *formula of reduction*.

From this, we have

$$I_1 = I - VD^2. \qquad (120)$$

EXAMPLES.

1. The moment of inertia of a rectangle in reference to one end as an axis being $\tfrac{1}{3}bd^3$, required the moment in reference to a parallel axis through the centre.

Equation (120) gives

$$I_1 = \tfrac{1}{3}bd^3 - bd\,(\tfrac{1}{2}d)^2 = \tfrac{1}{12}bd^3.$$

2. Given the moment of inertia of a triangle in reference to an axis through its vertex and parallel to the base, to find the moment relative to a parallel axis through its centre.

Example 7 of the preceding Article gives $I = \frac{1}{4}bd^3$; hence equation (120) gives

$$I_1 = \frac{1}{4}bd^3 - \frac{1}{2}bd\,(\tfrac{2}{3}d)^2 = \tfrac{1}{36}bd^3$$

3. Find the moment of the same triangle in reference to the base as an axis.

Equation (119) gives

$$I = \tfrac{1}{36}bd^3 + \tfrac{1}{2}bd\,(\tfrac{1}{3}d)^2 = \tfrac{1}{12}bd^3.$$

105. To find the relation between the moments of inertia in reference to different pairs of rectangular axes having the same origin.

Let x and y be rectangular axes, x_1 and y_1, also rectangular, having the same origin;
$a =$ the angle between x and x_1;
$I_x =$ the moment of inertia relatively to the axis x, similarly for
I_y, I_{x_1} and I_{y_1};
$B = \int xy\,dA$; and
$B_1 = \int x_1 y_1\,dA$.

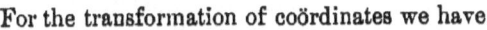

Fig. 99.

For the transformation of coördinates we have

$$x_1 = y \sin a + x \cos a;$$
$$y_1 = y \cos a - x \sin a;$$
$$x_1^2 + y_1^2 = x^2 + y^2.$$

Also
$$dA = dx\,dy = dx_1\,dy_1.$$

Hence,

$$\left.\begin{array}{l} I_{x_1} = \int y_1^2\,dA = I_x \cos^2 a + I_y \sin^2 a - 2B \cos a \sin a; \\ I_{y_1} = I_x \sin^2 a + I_y \cos^2 a + 2B \cos a \sin a; \\ B_1 = (I_x - I_y) \cos a \sin a + B (\cos^2 a - \sin^2 a); \\ \therefore\ I_{x_1} + I_{y_1} = I_x + I_y = I_p; \end{array}\right\} \quad (121)$$

the last value of which is found from the expression $\int y_1^2 dA + \int x_1^2 dA = \int (y_1^2 + x_1^2)\,dA = \int \rho^2 dA = I_p$; which shows that the *polar moment* equals the sum of two *rectangular moments*, the origin being the same. If the rectangular moments equal one another, we have $I_p = 2I_x$. Thus, in the circle, $I_p = \frac{1}{2}\pi r^4$. (See Ex. 4, Article 103), hence $I_x = \frac{1}{4}\pi r^4$.

The last of equations (121) is an *isotropic* function; since the sum of the moments relatively to a pair of rectangular axes, equals the sum of the moments relatively to any other pair of rectangular axes having the same origin; or, in other words, the sum of the moments of inertia relatively to a pair of rectangular axes, is constant.

To find the maximum or minimum moments we have, from the preceding equations,

$$\frac{dI_{x_1}}{da} = -(I_x - I_y)\cos a \sin a - B(\cos^2 a - \sin^2 a) = 0;$$

and

$$\frac{dI_{y_1}}{da} = +(I_x - I_y)\cos a \sin a + B(\cos^2 a - \sin^2 a) = 0;$$

$$\therefore B_1 = 0.$$

From the first or second of these we have

$$\frac{-2B}{I_x - I_y} = \frac{2\cos a \sin a}{\cos^2 a - \sin^2 a} = \tan 2a.$$

It may be shown by the ordinary tests that when I_{x_1} is a maximum, I_{y_1} will be a minimum, and the reverse; hence *there is always a pair of rectangular axes in reference to one of which the moment of inertia is greater than for any other axis, and for the other it is less.*

These are called *principal axes*.

Thus, in the case of a rectangle, if the axes are parallel to the sides and pass through the centre, we find

$$B = \iint_{-\frac{1}{2}d}^{+\frac{1}{2}d} xy\, dA = 0;$$

hence x and y are the axes for maximum and minimum moments; and if $d > b$, $\frac{1}{12}bd^3$ is the maximum, and $\frac{1}{12}b^3d$ a minimum moment of inertia for all axes passing through the origin. In a similar way we find that if the origin be at any other point the axes must be parallel to the sides for maximum and minimum moments.

The preceding analysis gives the position of the axes for maximum and minimum moments, when the moments are known in reference to any pair of rectangular axes. But if the axes for maximum and minimum moments are known as I_x and I_y, then $B = 0$; and calling these $I_{x'}$ and $I_{y'}$, Eqs. (121) become

$$\left.\begin{array}{l} I_{x_1} = I_{x'} \cos^2 a + I_{y'} \sin^2 a\,; \\ I_{y_1} = I_{x'} \sin^2 a + I_{y'} \cos^2 a\,; \\ B_1 = (I_{x'} - I_{y'}) \cos a \sin a. \end{array}\right\} \quad (122)$$

In the case of a square when the axes pass through the centre $I_x' = I_y'$;

$$\therefore I_{x_1} = I_{x'} (\cos^2 a + \sin^2 a) = I_{x'}\,;$$
$$I_{y_1} = I_{y'}, \text{ and}$$
$$B_1 = 0\,;$$

hence the moment of inertia of a square is the same in reference to all axes passing through its centre. The same is true for all regular polygons, and hence for the circle.

Examples.

1. To find the moment of inertia of a rectangle in reference to an axis through its centre and inclined at an angle a to one side, we have

$$I_x = \tfrac{1}{12} b d^3 \text{ and } I_y = \tfrac{1}{12} b^3 d$$
$$\therefore I_{x_1} = \tfrac{1}{12} b d (d^2 \cos^2 a + b^2 \sin^2 a)\,;$$
$$I_{y_1} = \tfrac{1}{12} b d (d^2 \sin^2 a + b^2 \cos^2 a).$$

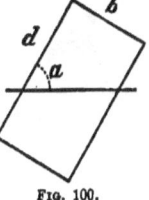

Fig. 100.

2. To find the moment of inertia of an isosceles triangle in reference to an axis through its centre and inclined at an angle a to its axis of symmetry.

We have $I_x = \tfrac{1}{36} b d^3$ and $I_y = \tfrac{1}{48} b^3 d$, in which b is the base and d the altitude;

$$\therefore I_x = \tfrac{1}{36} b d (d^2 \cos^2 a + \tfrac{3}{4} b^2 \sin^2 a)$$
$$I_y = \tfrac{1}{36} b d (d^2 \sin^2 a + \tfrac{3}{4} b^2 \cos^2 a).$$

The moment of inertia of a regular polygon about an axis

through its centre may be found by dividing it into triangles having their vertices at the centre of the polygon, and for bases the sides of the polygon; then finding the moments of the triangles about an axis through their centre and parallel to the given axis and reducing them to the given axis by the *formula of reduction*.

If R be the radius of the circumscribed circle, r that of the inscribed circle, and A the area of the polygon; then, for a regular polygon, we would find that

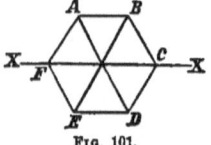

Fig. 101.

$$I = \tfrac{1}{12} A (R^2 + 2r^2).$$

For the circle $R = r$,

$$\therefore I = \tfrac{1}{4}\pi r^4,$$

as before found.

For the square, $r = \tfrac{1}{2}b$, $R = \tfrac{1}{2}b\sqrt{2}$, and $A = b^2$;

$$\therefore I = \tfrac{1}{12}b^4,$$

as before found.

106. Examples of the Moment of Inertia of Solids.

(The following results are taken from Mosley's *Mechanics and Engineering*.)

Fig. 102.

1. The moment of inertia of a solid cylinder about its axis of symmetry, r being its radius and h its height, is $\tfrac{1}{2}\pi h r^4$.

2. If the cylinder is hollow, c the thickness of the solid part and R the mean radius (equal to one-half the sum of the external and internal radii), then $I = 2\pi h c R (R^2 + \tfrac{1}{4}c^2)$.

3. The moment of inertia of a cylinder in reference to an axis passing through its centre and perpendicular to its axis of symmetry is $\tfrac{1}{4}\pi h r^2 (r^2 + \tfrac{1}{3}h^2)$.

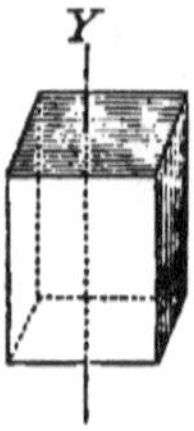

4. The moment of inertia of a rectangular parallelopipedon about an axis passing through its centre and parallel to one of its edges. Let a be the length of the edge parallel to the axis, and b and c the lengths of the other edges, then $I = \tfrac{1}{12} abc (b^2 + c^2) = \tfrac{1}{12}$ *of the volume multiplied by the square of the diagonal of the base.*

Fig. 103.

5. The moment of inertia of an upright triangular prism having an isosceles triangle for its base, in reference to a vertical axis passing through its centre of gravity.

Let the base of the triangle be a, its altitude b, and the altitude of the prism be h, then

$$I = \tfrac{1}{12}abh\,(\tfrac{1}{4}a^2 + \tfrac{1}{9}b^2).$$

FIG. 104.

6. The moment of inertia of a cone in reference to an axis of symmetry is $\tfrac{1}{10}\pi r^4 h$. (r being the radius of the base and h the altitude.)

 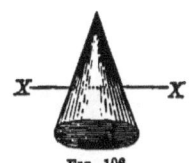 

FIG. 105.   FIG. 106.   FIG. 107.

7. The moment of inertia of a cone in reference to an axis through its centre and perpendicular to its axis of symmetry is $\tfrac{1}{20}\pi r^2 h\,(r^2 + \tfrac{1}{3}h^2).$

8. The moment of inertia of a sphere about one of its diameters is $\tfrac{8}{15}\pi R^5$.

9. The moment of inertia of a segment of a sphere about a diameter parallel to the plane of section.

Let R be the radius of the sphere, and b the distance of the plane section from the centre, then

$$I = \tfrac{1}{60}\pi\,(16R^5 + 15R^4 b + 10R^2 b^3 - 9b^5).$$

FIG. 108.

RADIUS OF GYRATION.

107. We may conceive the mass to be concentrated at such a point that the moment of inertia in reference to any axis will be the same as for the distributed mass in reference to the same axis.

The radius of gyration is the distance from the moment axis to a point in which, if the entire mass be concentrated, the moment of inertia will be the same as for the distributed mass

The principal radius of gyration is the radius of gyration in reference to a moment axis through the centre of the mass.

Let k = the radius of gyration;
k_1 = the principal radius of gyration;
M = the mass of the body; and
D = the distance between parallel axes;

then, according to the definitions and equation (119), we have

$$Mk^2 = \Sigma mr^2$$
$$= \Sigma mr_1^2 + MD^2$$
$$= Mk_1^2 + MD^2;$$
$$\therefore k^2 = k_1^2 + D^2; \qquad (123)$$

from which it appears that k is a minimum, for $D = 0$, in which case $k = k_1$; that is, the *principal radius of gyration is the minimum radius for parallel axes*.

We have
$$I_1 = Mk_1^2;$$
$$\therefore k_1^2 = \frac{I_1}{M};$$

hence, the square of the principal radius of gyration equals the moment of inertia in reference to a moment axis through the centre of the body divided by the mass.

EXAMPLES.

1. Find the principal radius of gyration of a circle in reference to a rectangular axis.

Example 5 of Article 103 gives, $I_1 = \frac{1}{4}\pi r^4$, which is the moment of an area, hence, we use πr^2 for M, and have

$$k_1^2 = \frac{\frac{1}{4}\pi r^4}{\pi r^2} = \frac{1}{4}r^2.$$

2. For a circle in reference to a polar axis, $k_1^2 = \frac{1}{2}r^2$.

3. For a straight line in reference to a moment axis perpendicular to it, $k_1^2 = \frac{1}{12}l^2$.

4. For a sphere, $k_1^2 = \frac{2}{5}r^2$.

5. For a rectangle whose sides are respectively a and b, in reference to an axis perpendicular to its plane, $k_1^2 = \frac{1}{12}(a^2 + b^2)$.

6. Find the principal radius of gyration of a cone when the moment axis is the axis of symmetry.

CHAPTER VIII.

MOTION OF A PARTICLE FREE TO MOVE IN ANY DIRECTION.

108. A free, material particle, acted upon by a system of forces which are not in equilibrium among themselves, will describe a path which will be a straight or a curved line.

The direction of motion at any point of the path will coincide with that of the action-line of the resultant of all the forces which have been impressed upon the particle prior to reaching the point, which will also coincide with the tangent to the path at that point.

Let ds be an element of the path described by the particle in an element of time dt; R the resultant of the impressed forces, and m the mass of the particle; then, according to Article 21, we have

$$R - m\frac{d^2s}{dt^2} = 0.$$

Let a be the angle between the action-line of the resultant R (or of the arc ds) and the axis of x; multiplying by $\cos a$, we have

$$R \cos a - m\frac{d^2s}{dt^2}\cos a = 0;$$

in which $R \cos a$ is the *x-component* of the resultant, and according to equation (51) equals X; or, in other words, it is the projection of the line representing the resultant on the axis of x; $d^2s \cos a$ is the projection of d^2s on the axis of x, and is d^2x. Hence, the equation becomes,

and similarly,
$$\left.\begin{aligned} X - m\frac{d^2x}{dt^2} &= 0; \\ Y - m\frac{d^2y}{dt^2} &= 0; \\ Z - m\frac{d^2z}{dt^2} &= 0; \end{aligned}\right\} \quad (124)$$

which are the equations for the motion of a particle along the coördinate axes; and are also the equations for the motion of a body of finite size when the action-line of the resultant passes through the centre of the mass. They are also the *equations of translation* of the centre of any free mass when the forces produce both rotation and translation; in which case m should be changed to M to represent the total mass. See Article 38.

VELOCITY AND LIVING FORCE.

109. Multiplying the first of equations (124) by dx, the second by dy, and the third by dz, adding and reducing, give

$$Xdx + Ydy + Zdz = \tfrac{1}{2}md\left(\frac{dx^2 + dy^2 + dz^2}{dt^2}\right) = \tfrac{1}{2}md\frac{ds^2}{dt^2};$$

and integrating gives

$$\int(Xdx + Ydy + Zdz) = \tfrac{1}{2}m\frac{ds^2}{dt^2} = \tfrac{1}{2}mv^2 + C.$$

The first member is the work done by the impressed forces; for if R be the resultant, and s the path, then, according to Article 25, equation (26), the work will be $\int Rds$, and by projecting this on the coördinate axes and taking their sum, we have the above expression. The second member is the stored energy plus a constant.

Let X, Y, Z be known functions of x, y, z, and that the terms are integrable. (It may be shown that they are always integrable when the forces act towards or from fixed centres.) Performing the integration between the limits x_0, y_0, z_0, and x_1, y_1, z_1, we have

$$\phi(x_0, y_0, z_0) - \phi(x_1, y_1, z_1) = \tfrac{1}{2}m(v^2_0 - v^2_1); \quad (125)$$

hence, the work done by the impressed forces upon a body in passing from one point to another equals the difference of the living forces at those points. It also appears that the velocity at two points will be independent of the path described; also, that, when the body arrives at the initial point, it will have the same velocity and the same energy that it previously had at that point.

Examples.

1. *If a body is projected into space, and acted upon only by gravity and the impulse; required the curve described by the projectile.*

Take the coördinate plane xy in the plane of the forces, x horizontal and y vertical, the origin being at the point from which the body is projected.

Let $W=$ the weight of the body;
$v=$ the velocity of projection; and
$a = BAX =$ the angle of elevation at which the projection is made.

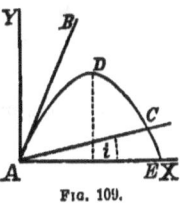

Fig. 109.

We have,
$$X = 0; \quad Y = -mg; \quad Z = 0; \quad z = 0;$$
and equations (124) become
$$\frac{d^2x}{dt^2} = 0;$$
$$-g - \frac{d^2y}{dt^2} = 0.$$

Integrating, observing that $v \cos a$ will be the initial velocity along the axis of x, and $v \sin a$ that along y, we have,
$$\frac{dx}{dt} = v \cos a;$$
$$\frac{dy}{dt} = v \sin a - gt;$$
and integrating again, observing that the initial spaces are zero, we have,
$$\left. \begin{array}{l} x = vt \cos a; \\ y = vt \sin a - \tfrac{1}{2}gt^2. \end{array} \right\} \qquad (a)$$

Eliminating t from these equations, gives
$$y = x \tan a - \frac{gx^2}{2v^2 \cos^2 a}; \qquad (b)$$

which is the equation of the common parabola, whose axis is parallel to the axis of y.

12

Let h be the height through which a body must fall to acquire a velocity v, then $v^2 = 2gh$, and the equation (b) becomes,

$$y = x \tan a - \frac{x^2}{4h \cos^2 a}. \qquad (c)$$

To find the range AE,
make $y = 0$ in equation (b), and we find $x = 0$, and

$$x = AE = 4h \cos a \sin a = 2h \sin 2a; \qquad (d)$$

which is a maximum for $a = 45°$. The range will be the same for two angles of elevation, one of which is the complement of the other.

The greatest height,
will correspond to $x = h \sin 2a$, which, substituted in (b), gives,

$$h \sin^2 a. \qquad (e)$$

The velocity at the end of the time t
is,

$$V = \frac{ds}{dt} = \sqrt{\left\{\frac{dx^2}{dt^2} + \frac{dy^2}{dt^2}\right\}} = \sqrt{v^2 - 2v g t \sin a + g^2 t^2}; \qquad (f)$$

or by eliminating t by means of the first of equations (a), we have,

$$V = \sqrt{v^2 - 2gx \tan a + \frac{gx^2}{2h \cos^2 a}}. \qquad (g)$$

The direction of motion at any point
is found by differentiating equation (c), and making

$$\tan \theta = \frac{dy}{dx} = \tan a - \frac{x}{2h \cos a}. \qquad (h)$$

At the highest point $\theta = 0$, $\therefore x = h \sin 2a$, as before found. For $x = 2h \sin 2a$, we have,

$$\tan \theta = - \tan a,$$

or the angle at the end of the range is the supplement of the angle of projection.

2. A body is projected at an angle of elevation of 45°, and has a range of 1,000 feet; required the velocity of projection, the time of flight, and the parameter of the parabola.

3. What must be the angle of elevation in order that the horizontal range may equal the greatest altitude? What, that it may equal n times the greatest altitude?

4. Find the velocity and the angle of elevation of a projectile, so that it may pass through the points whose coördinates are $x_1 = 400$ feet, $y_1 = 50$ feet, $x_2 = 600$ feet, and $y_2 = 40$ feet.

5. If the velocity is 500 feet per second, and the angle of elevation 45 degrees; required the range, the greatest elevation, the velocity at the highest point, the direction of motion 6,000 feet from the point of projection, and the velocity at that point.

6. If a plane, whose angle of elevation is i, passes through the origin, find the coördinates of the point C, Fig. 109, where the projectile passes it.

7. In the preceding problem, if i is an angle of depression, find the coördinates.

8. Find the equation of the path when the body is projected horizontally.

9. If a body is projected in a due southerly direction at an angle of elevation a, and is subjected to a constant, uniform, horizontal pressure in a due easterly direction; required the equations of the path, neglecting the resistance of the air.

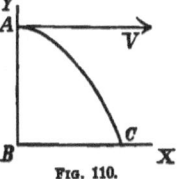

Fig. 110.

We have

$$X = 0; \quad Y = -mg; \quad Z = F \text{ (a constant)}.$$

The projection of the path on the plane xy will be a parabola, on xz also a parabola.

10. If a body is projected into the air, and the resistance of the air varies as the square of the velocity; required the equation of the curve.

(The final integrals for this problem cannot be found. Approximate solutions have been made for the purpose of determining certain laws in regard to gunnery. It is desirable for the student to establish the equations and make the first steps in the reduction.)

(The remainder of this chapter may be omitted without detriment to what follows it. It, however, contains an interesting topic in Mechanics, and is of vital importance in Mathematical Astronomy and Physics.)

CENTRAL FORCES.

110. Central forces are such as act directly towards or from a point called a centre. Those which act towards the centre are called *attractive*, and are considered negative, while those which act from the centre are *repulsive*, and are considered positive. The centre may be fixed or movable.

The line from the centre to the particle is called a *radius vector*. The path of a body under the action of central forces is called an orbit.

The forces considered in Astronomy and many of those in Physics, are central forces.

GENERAL EQUATIONS.

111. Consider the force as attractive, and let it be represented by $-F$.

Take the coördinate plane xy in the plane of the orbit, the origin being at the centre of the force, and $OP = r$, the radius vector, then

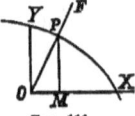

FIG. 111.

$$X = -F\cos\alpha = -F\frac{x}{r};$$

$$Y = -F\cos\beta = -F\frac{y}{r};$$

and the first two of equations (124) become

$$\left. \begin{array}{l} m\dfrac{d^2x}{dt^2} = -F\dfrac{x}{r}; \\[1em] m\dfrac{d^2y}{dt^2} = -F\dfrac{y}{r}. \end{array} \right\} \qquad (126)$$

To change these to polar coördinates, first modify them by multiplying the first by y and the second by x, and subtracting, and we have

$$my\frac{d^2x}{dt^2} - mx\frac{d^2y}{dt^2} = 0;$$

and multiplying the first by x and the second by y, and adding, we have

$$mx\frac{d^2x}{dt^2} + my\frac{d^2y}{dt^2} = -Fr.$$

Let $\theta = POM =$ the variable angle, then

$$x = r\cos\theta, \qquad y = r\sin\theta,$$

and differentiating each twice, we find

$$d^2x = (d^2r - rd\theta^2)\cos\theta - (2drd\theta + rd^2\theta)\sin\theta;$$

$$d^2y = (d^2r - rd\theta^2)\sin\theta + (2drd\theta + rd^2\theta)\cos\theta;$$

which substituted in the preceding equations, give

$$\frac{d^2r}{dt^2} - r\left(\frac{d\theta}{dt}\right)^2 = -\frac{F}{m}; \qquad (127)$$

$$2\frac{drd\theta}{dt^2} + r\frac{d^2\theta}{dt^2} = 0,$$

which may be put under the form

$$\frac{1}{r}\frac{d}{dt}\left(r^2\frac{d\theta}{dt}\right) = 0. \qquad (128)$$

Equation (127) shows that the acceleration along r is the force on a unit of mass; and (128) shows that there is no acceleration perpendicular to the radius vector.

112. *Principle of equal areas.*—Integrate equation (128), and we have

$$r^2\frac{d\theta}{dt} = C; \qquad (129)$$

and integrating a second time, we have

$$\int r^2 d\theta = Ct; \qquad (130)$$

the constant of integration being zero, since the initial values of t and θ are both zero. But from Calculus $\int r^2 d\theta$ is twice the sectoral area POX; hence the sectoral area swept over by the radius vector increases directly as the time; and *equal areas will be passed over in equal times.*

Making $t = 1$, we find that C will be twice the sectoral area passed over in a unit of time.

The converse is also true, *that if the areas are proportional to the times the force will be central.*

For, multiplying the first of (126) by y and the second by x and taking their difference, we have

$$m\frac{d^2x}{dt^2}y - m\frac{d^2y}{dt^2}x = 0;$$

or,

$$Xy - Yx = 0;$$

which is the equation of a straight line, and is the equation of the action-line of the resultant, and since it has no absolute term it passes through the origin.

113. *To find the equation of the orbit,* eliminate dt from equations (127) and (128). For the sake of simplifying the final equation, make $r = \dfrac{1}{u}$, and (129) becomes

$$\frac{1}{u^2} = C\frac{dt}{d\theta}. \qquad (131)$$

Differentiating and reducing, gives

$$dr = -\frac{du}{u^2} = -C\frac{du\,dt}{d\theta};$$

or,

$$\frac{dr}{dt} = -C\frac{du}{d\theta}.$$

the first member of which is the velocity in the direction of the radius.

Differentiating again, gives

$$\frac{d^2r}{dt^2} = -C^2u^2\frac{d^2u}{d\theta^2}.$$

Substituting these in equation (127) and at the same time making m equal to unity, since the unit is arbitrary, we have

$$-C^2u^2\frac{d^2u}{d\theta^2} - C^2u^3 = -F,$$

or,
$$\frac{d^2u}{d\theta^2} + u - \frac{F}{C^2u^2} = 0; \qquad (132)$$

which is the differential equation of the orbit.

When the law of the force is known, the value of F may be substituted, and the equation integrated, and the orbit be definitely determined.

Multiplying by du and integrating the first two terms, we have
$$C^2\left(\frac{du^2}{d\theta^2} + u^2\right) - 2\int F\frac{du}{u^2} = C_1. \qquad (133)$$

114. *Given the equation of the orbit, to find the law of the force.*

From equation (132), we have
$$F = C^2u^2\left(\frac{d^2u}{d\theta^2} + u\right). \qquad (134)$$

Another expression is deduced as follows: let

$p =$ the perpendicular from the centre on the tangent, then from Calculus we have

$$p^2 = \frac{r^4 d\theta^2}{ds^2} = \frac{r^4 d\theta^2}{dr^2 + r^2 d\theta^2}$$

$$= \frac{1}{\dfrac{du^2}{d\theta^2} + u^2}. \qquad (135)$$

Differentiating, gives
$$p\,dp = -\frac{\dfrac{d^2u}{d\theta^2} + u}{\left(\dfrac{du^2}{d\theta^2} + u^2\right)^2}du;$$

and dividing by p^4, substituting, $du = -u^2 dr$, and reducing, we have

$$\frac{1}{p^3} \cdot \frac{dp}{dr} = u^2 \left(\frac{d^2 u}{d\theta^2} + u\right);$$

which, combined with equation (134), gives

$$F = \frac{C^2}{p^3} \cdot \frac{dp}{dr}; \qquad (136)$$

which is a more simple formula for determining the law of the central force.

115. *To determine the velocity at any point of the orbit.*
We have

$$v = \frac{ds}{dt} = \frac{ds}{dt}\frac{d\theta}{d\theta} = \frac{ds}{d\theta}\frac{d\theta}{dt}$$

$$= Cu^2 \frac{ds}{d\theta} \text{ (from equation (131))}$$

$$= \frac{C}{p}. \text{ (from } Dif.\ Cal.), \qquad (137)$$

Hence, *the velocity varies inversely as the perpendicular from the centre upon the tangent to the orbit.*

Another expression is found by substituting the value of p, equation (135), in (137).

Hence,

$$v = C\sqrt{\frac{du^2}{d\theta^2} + u^2}. \qquad (138)$$

Still another expression may be found by substituting equation (138) in (133); hence

$$v^2 = C_1 + 2\int F \frac{du}{u^3} \qquad (139)$$

$$= C_1 - 2\int F\, dr.$$

Since F is a function of r, the integral of this equation gives v in terms of r, or the velocity depends directly upon the distance of the body from the centre. Hence, the velocity at any two points in the orbit is independent of the path between them, the *law* of the force remainining the same.

116. *To determine the time of describing any portion of the orbit.*

To find it in terms of r, eliminate $d\theta$ between equations (129) and (133), reduce and find

$$t = \int_{r_1}^{r_2} \frac{dr}{\sqrt{C_1 - \frac{C^2}{r^2} - 2\int F dr}};\qquad (140)$$

which integrated gives the time.

To find it in terms of the angle, we have from (129)

$$t = \frac{1}{C}\int r^2 d\theta;$$

from which r must be eliminated by means of the equation of the orbit, and the integration performed in reference to θ.

117. *To find the components of the force along the tangent and normal.*

Let $T =$ the tangential component;
$N =$ the normal component;
and resolving them parallel to x and y, we have

$$m\frac{d^2x}{dt^2} = X = T\frac{dx}{ds} - N\frac{dy}{ds},$$

$$m\frac{d^2y}{dt^2} = Y = T\frac{dy}{ds} + N\frac{dx}{ds}.$$

Eliminating N gives

$$m\frac{d^2x}{dt^2}dx + m\frac{d^2y}{dt^2}dy = Tds;$$

or

$$T = X\frac{dx}{ds} + Y\frac{dy}{ds}, \text{ also } = m\frac{d^2s}{dt^2}.$$

Eliminating T gives

$$N = m\frac{dx}{ds}\frac{d^2y}{dt^2} - m\frac{dy}{ds}\frac{d^2x}{dt^2}$$

$$= \frac{mds^3}{dt^2 ds}\left(\frac{dxd^2y - dyd^2x}{ds^3}\right)$$

$$= m\frac{v^2}{\rho}; \qquad (141)$$

hence *the component of the force in the direction of the normal is dependent entirely upon the velocity and radius of curvature*. This is called the *centrifugal force*. It is the measure of the force which deflects the body from the tangent. The force directed towards the centre is called *centripetal*. (It is attractive). When the normal coincides with the radius vector the centrifugal force is *directly* opposed to the centripetal.

If $\omega =$ the angular velocity described by the radius of curvature, then, $v = \rho\omega$, and equation (141) becomes

$$N = m\omega^2\rho. \qquad (142)$$

EXAMPLES.

1. If a body on a smooth, horizontal plane is fastened to a point in the plane by means of a string, what will be the number of revolutions per minute, that the tension of the string may be twice the weight of the body.

2. A body whose weight is ten pounds, revolves in a horizontal circle whose radius is five feet, with a velocity of forty feet per second; required the tension of the string which holds it. (Use equation (141).)

3. Required the velocity and periodic time of a body revolving in a circle at a distance of n radii from the earth's centre.

The weight of the body on the surface being mg, at the distance of n radii it will weigh $mg\left(\dfrac{r}{nr}\right)^2 = \dfrac{mg}{n^2}$, and this is a measure of the force at that distance. (Use equation (141) or (139).)

$$\text{Ans. } v = \left(\frac{gr}{n}\right)^{\frac{1}{2}}; \quad t = \left(\frac{n^3 r}{g}\right)^{\frac{1}{2}}.$$

(This is substantially the problem which Sir Isaac Newton used to prove the law of Universal Gravitation. See Whewell's *Inductive Sciences*.)

4. *A particle is projected from a given point in a given direction with a given velocity, and moves under the action of a force which varies inversely as the square of the distance from the centre; required the orbit.*

Let μ = the force at a unit's distance, then

$$F = \mu u^2,$$

and equation (132) becomes

$$\frac{d^2 u}{d\theta^2} + u - \frac{\mu}{C^2} = 0;$$

or,

$$\frac{d^2}{d\theta^2}\left(u - \frac{\mu}{C^2}\right) = -\left(u - \frac{\mu}{C^2}\right);$$

the first integral of which becomes by reduction

$$d\theta = \frac{-d\left(u - \frac{\mu}{C^2}\right)}{\sqrt{A - \left(u - \frac{\mu}{C^2}\right)^2}};$$

in which A is an arbitrary constant, and the negative value of the radical is used.

Integrating again, making θ_0 the arbitrary constant, we have

$$\theta - \theta_0 = \cos^{-1}\left\{\frac{u - \frac{\mu}{C^2}}{A}\right\};$$

which by reduction gives

$$u = \frac{1}{r} = \frac{\mu}{C^2}\left(1 + \frac{AC^2}{\mu}\cos(\theta - \theta_0)\right), \qquad (a)$$

which is the general polar equation of a conic section, the origin being at the focus. As this is the law of Universal Gravitation, it follows that the orbits of the planets and comets are conic sections having the centre of the sun for the focus. In equation (a), θ_0 is the angle between the major axis and a line drawn through the centre of the force, and $\dfrac{AC^2}{\mu}$ is the eccentricity $= e$; hence the equation may be written

$$u = \frac{\mu}{C^2}\left(1 + e \cos(\theta - \theta_0)\right). \qquad (b)$$

The magnitude and position of the orbit will be determined from the constants which enter the equation, and these are determined by knowing the position, velocity, and direction of motion at some point in the orbit.

Draw a figure to represent the orbit, and make a tangent to the curve at a point which we will consider the initial point. Let β be the angle between the path and the radius vector at the initial point, r_0 the initial radius vector, and V_0 the initial velocity; then at the initial point

$$u = \frac{1}{r_0}, \quad \theta = 0, \quad \cot\beta = \frac{dr}{r_0 d\theta} = -\frac{du}{u d\theta}, \qquad (c)$$

and from equation (b)

$$e \cos \theta_0 = \frac{C^2}{\mu r_0} - 1, \qquad (d)$$

$$\frac{du}{d\theta} = \frac{\mu e}{C^2}\sin\theta_0,$$

which, combined with equation (c), gives

$$\frac{C^2 \cot\beta}{\mu r_0} = -e \sin\theta_0. \qquad (e)$$

From equation (137)

$$C = V_0 r_0 \sin\beta; \qquad (f)$$

which, substituted in equations (d) and (e), and the latter divided by the former, gives

$$\tan \theta_0 = \frac{V_0^2 r_0 \sin \beta \cos \beta}{\mu - V_0^2 r_0 \sin^2 \beta}.$$

Squaring equations (d) and (e), adding and reducing by equation (f), give

$$e^2 = 1 - \frac{V_0^2 r_0^2 \sin^2 \beta}{\mu}\left(\frac{2}{r_0} - \frac{V_0^2}{\mu}\right). \quad (g)$$

Hence, when

$V_0^2 > \dfrac{2\mu}{r_0}$, $e > 1$, and the orbit is a hyperbola,

$V_0^2 = \dfrac{2\mu}{r_0}$, $e = 1$, and the orbit is a parabola,

$V_0^2 < \dfrac{2\mu}{r_0}$, $e < 1$, and the orbit is an ellipse.

or (see example 23, page 34) the orbit will be a hyperbola, a parabola, or an ellipse, according as the velocity of projection is greater than, equal to, or less than the velocity from infinity.

As the result of a large number of observations upon the planets, especially upon Mars, Kepler deduced the following laws:

1. The planets describe ellipses of which the Sun occupies a focus.

2. The radius vector of each planet passes over equal areas in equal times.

3. The squares of the periodic times of any two planets are as the cubes of the major axes of their orbits.

The first of these is proved by the preceding problem, since the orbits are reënterant curves. The second is proved by equation (130). The third we will now prove.

5. *Required the relation between the time of a complete circuit of a particle in an ellipse, and the major axis of the orbit.*

Let the initial point be at the extremity of the major axis near the pole, then θ_0 in equation (d) will be zero, and we have

$$C^2 = \mu r_0 (1 + e);$$

but from the ellipse,
$$r_0 = a - ae = a(1-e);$$
$$\therefore C = \sqrt{\mu a(1-e^2)}. \qquad (a)$$

Equation (130) gives
$$T = \frac{2 \text{ area of ellipse}}{C}$$
$$= \frac{2\pi a^2 \sqrt{(1-e^2)}}{\sqrt{\mu a(1-e^2)}}$$
$$= 2\pi \sqrt{\frac{a^3}{\mu}};$$
$$\therefore T^2 \propto a^3.$$

6. The orbit being an ellipse, required the law of the force
The polar equation of the ellipse, the pole being at the focus, is
$$u = \frac{1}{r} = \frac{1 + e \cos(\theta - \theta_0)}{a(1-e^2)};$$

which, differentiated twice, gives
$$\frac{d^2u}{d\theta^2} = -\frac{e \cos(\theta - \theta_0)}{a(1-e^2)};$$

and these, in equation (132), give,
$$F = \frac{C^2}{a(1-e^2)} \frac{1}{r^2};$$

hence *the force varies inversely as the square of the distance.*

7. Find the law of force by which the particle may describe a circle, the centre of the force being in the circumference of the circle. (Tait and Steele, *Dynamics of a Particle.*)

$$\text{Ans. } F \propto \frac{1}{r^5}.$$

8. If the force varies directly as the distance, and is attractive, determine the orbit.

(This is the law of molecular action, and analysis shows that the orbit is an ellipse. The problem is of great importance in Physics, especially in Optics and Acoustics.)

CHAPTER IX.

CONSTRAINED MOTION OF A PARTICLE.

118. If a body is compelled to move along a given fixed curve or surface, it is said to be constrained. The given curve or surface will be subjected to a certain pressure which will be normal to it.

If instead of the curve or surface, a force be substituted for the pressure which will be continually normal to the surface, and whose intensity will be exactly equal and opposite to the pressure on the curve, the particle will describe the same path as that of the curve, and the problem may be treated as if the particle were free to move under the action of this system of forces.

Let $N =$ the normal pressure on the surface, and
$L = f(x, y, z) = 0$, be the equation of the surface;
$\theta_x, \theta_y, \theta_z$, the angles between N and the respective coördinate axes.

Then

$$\left. \begin{array}{l} X + N \cos \theta_x - m \dfrac{d^2x}{dt^2} = 0 \, ; \\[4pt] Y + N \cos \theta_y - m \dfrac{d^2y}{dt^2} = 0 \, ; \\[4pt] Z + N \cos \theta_z - m \dfrac{d^2z}{dt^2} = 0 \, ; \end{array} \right\} \quad (143)$$

in which the third terms are the measures of the resultants of the axial components of the applied forces. We will confine the further discussion to forces in a plane. Take xy in the plane of the forces, then we have

$$\left. \begin{array}{l} m \dfrac{d^2x}{dt^2} = X - N \dfrac{dy}{ds} \, ; \\[4pt] m \dfrac{d^2y}{dt^2} = Y + N \dfrac{dx}{ds} \, . \end{array} \right\} \quad (144)$$

Eliminating N, we find

$$m\left\{\frac{dxd^2x}{dt^2} + \frac{dyd^2y}{dt^2}\right\} = Xdx + Ydy,$$

or

$$m\left\{d\left(\frac{dx}{dt}\right)^2 + d\left(\frac{dy}{dt}\right)^2\right\} = 2Xdx + 2Ydy;$$

and integrating, making v_0 the initial velocity along the path, and v the velocity at any other point, we have

$$\tfrac{1}{2}m(v^2 - v_0^2) = \int(Xdx + Ydy); \qquad (145)$$

hence, *the living force gained or lost in passing from one point to another is equal to the work done by the impressed forces.* If the forces X and Y are functions of the coördinates x and y, and the terms within the parenthesis are integrable, the result may be expressed in terms of constants and functions of the coördinates of the initial and terminal points, and may be written $\quad \tfrac{1}{2}m(v^2 - v_0^2) = c\,\phi(x_0,y_0) - c\,\phi(x,y);$
hence, for such a system, *the velocity will be independent of the path described, and will be dependent only upon the coördinates of the points;* also, *the velocity will be independent of the normal pressure.*

119. *To find the normal pressure*
multiply the first of equations (144) by dy, the second by dx, subtract, observing that $dx^2 + dy^2 = ds^2$, and we find

$$m\left\{\frac{dx}{ds}\frac{d^2y}{dt^2} - \frac{dy}{ds}\frac{d^2x}{dt^2}\right\} = Y\frac{dx}{ds} - X\frac{dy}{ds} + N;$$

or

$$m\frac{ds^2}{dt^2}\left\{\frac{dxd^2y - dyd^2x}{ds^3}\right\} = Y\frac{dx}{ds} - X\frac{dy}{ds} + N;$$

$$\therefore N = X\frac{dy}{ds} - Y\frac{dx}{ds} + m\frac{v^2}{\rho}; \qquad (146)$$

in which ρ is the radius of curvature at the point. The first and second terms of the second member are the normal components of the impressed forces. *The total normal pressure* will, therefore, be that due to the impressed forces plus that due to the force necessary to deflect the body from the tangent. The last term is called the *centrifugal force*, as stated in Article 117. If the body moves on the convex side of the curve, the last term should be subtracted from the others; hence it might be written $\pm m \dfrac{v^2}{\rho}$; in which $+$ belongs to movement on the concave arc, and $-$ on the convex.

120. *To find the time of movement,* from equation (145), we have

$$\frac{ds^2}{dt^2} - v_0^2 = \frac{2}{m} \int (X dx + Y dy) ;$$

$$\therefore t = \int_{s_0}^{s} \frac{\sqrt{m}\, ds}{\sqrt{2\int(X dx + Y dy) + m v_0^2}}. \quad (147)$$

121. *To find where the particle will leave the constraining curve.*

At that point $N = 0$, which gives

$$m \frac{v^2}{\rho} = X \frac{dy}{ds} - Y \frac{dx}{ds} ; \quad (148)$$

which, combined with the equation of the curve, makes known the point.

If a body is subjected only to the force of gravity, we have $X = 0$ in all the preceding equations.

EXAMPLES.

1. *A body slides down a smooth inclined plane under the force of gravity; required the formulas for the motion.*

Take the origin at the upper end and let the equation of the plane be

$$y = ax;$$

13

y being positive downward. Then we have
$$X = 0, \qquad Y = mg, \qquad dy = adx, \qquad v_0 = 0,$$
and equation (145) becomes
$$v^2 = 2fgadx = 2gax = 2gy ; \qquad (a)$$
hence, the velocity is the same as if it fell *vertically* through the same height.

To find the time, equation (147) gives
$$t = \int \frac{ds}{\sqrt{2gax}} = \sqrt{\frac{1+a^2}{2ga}} \int \frac{dx}{\sqrt{x}} = s\sqrt{\frac{2}{gy}} ; \qquad (b)$$
that is, *if the altitude of the plane (y) is constant the time varies directly as the length, s.*

We may also find
$$s = t\sqrt{\tfrac{1}{2}gy} = \tfrac{1}{2}gt^2 \sin a. \qquad (c)$$

2. Prove that the times of descent down all chords of a vertical circle which pass through either extremity of a vertical diameter are the same.

3. Find the straight line from a given point to a given inclined plane, down which a body will descend in the least time.

4. The time of descent down an inclined plane is twice that down its height; required the inclination of the plane to the horizon.

5. At the instant a body begins to descend an inclined plane, another body is projected up it with a velocity equal to the velocity which the first body will have when it reaches the foot of the plane; required the point where they will meet.

6. Two bodies slide down two inclined lines from two given points in the same vertical line to any point in a curve in the same time, the lines all being in one vertical plane; required the equation of the curve.

7. A given weight, P, draws another weight, W, up an inclined plane, by means of a cord parallel to the plane; through what distance must P act so that the weight, W, will move s feet after P is separated from it.

[121.]   SIMPLE PENDULUM.   195

8. *Required a curve such that if it revolve with a uniform angular velocity about a vertical diameter, and a smooth ring of infinitesimal diameter be placed upon it at any point, it will not slide on the curve.*

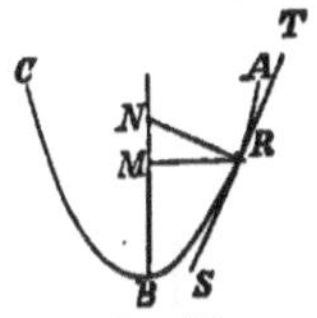

Let ω be the angular velocity, then we have

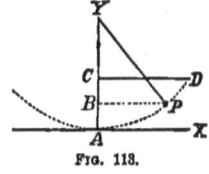
Fig. 112.

$$Y = -mg, \quad X = m\omega^2 x, \quad v = 0, \quad v_0 = 0,$$

and equation (145) becomes

$$\omega^2 x^2 - 2gy + C = 0,$$

which is the equation of the common parabola.
If the origin be taken at B, C will be zero.

(ANOTHER SOLUTION.—Let NR be a normal to the curve, MR = the centrifugal force, NM = the force of gravity; but the latter is constant, hence NM, the subnormal, is constant, which is a property of the common parabola.)

9. Find the normal pressure against the curve in the preceding problem.

10. THE PENDULUM.—*Find the time of oscillation of the simple pendulum.* This is equivalent to finding the time of descent of a particle down a smooth arc of a vertical circle.

Take the origin of coördinates at A, the lowest point. Let the particle start at D, at a height $AC = h$; when it has arrived at P, it will have fallen through a height $CB = h - y$, and, according to equation (a) on the preceding page, will have a velocity

$$v = \sqrt{2g(h-y)} = \frac{ds}{dt} \qquad (a)$$

The equation of the arc is

$$x^2 = 2ry - y^2;$$

hence

$$dx^2 = \frac{(r-y)^2}{2ry - y^2} dy^2.$$

Fig. 113.

But

$$ds^2 = dx^2 + dy^2,$$

$$\therefore ds = \frac{r\,dy}{\sqrt{2ry - y^2}}.$$

Considering this as negative, since for the descent the arc is a decreasing function of the time, we have from (a)

$$t = \frac{r}{\sqrt{2g}} \int_0^h \frac{dy}{\sqrt{(h-y)(2ry - y^2)}}$$

This may be put in a form for integration by Elliptic Functions; but by developing it into a series, each term may be easily integrated. In this way we find

$$t = \tfrac{1}{2}\pi\sqrt{\frac{r}{g}} \left\{ 1 + \left(\tfrac{1}{2}\right)^2 \frac{h}{2r} + \left(\tfrac{1\cdot 3}{2\cdot 4}\right)^2 \left(\frac{h}{2r}\right)^2 + \left(\tfrac{1\cdot 3\cdot 5}{2\cdot 4\cdot 6}\right)^2 \left(\frac{h}{2r}\right)^3 + \text{etc.} \right\};$$

by means of which the time may be approximated to, with any degree of accuracy.

When the arc is very small, all the terms containing $\dfrac{h}{2r}$ will be small, and by neglecting them, we have for a complete oscillation (letting l be the length of the pendulum),

$$T = 2t = \pi\sqrt{\frac{l}{g}}; \qquad (b)$$

that is, *for very small arcs the oscillations may be regarded as isochronal*, or performed in the same time.

For the same place *the times of vibration are directly as the square roots of the lengths of the pendulums*.

For any pendulum *the times of vibration vary inversely as the square roots of the force of gravity at different places*.

If t is constant

$$l \propto g.$$

11. What is the length of a pendulum which will vibrate three times in a second?

12. Prove that the lengths of pendulums vibrating during the

same time at the same place, are inversely as the square of the number of vibrations.

13. Find the time of descent of a particle down the arc of a cycloid.

The differential equation of the curve referred to the vertex as an origin, x being horizontal and y vertical (r being the radius of the generating circle), is

$$dx = \frac{2r - y}{\sqrt{2ry - y^2}} dy.$$

$$\text{Ans. } \pi \sqrt{\frac{r}{g}}.$$

The time will be the same from whatever point of the curve the motion begins, and hence, it is called *tautochronal*.

14. In the simple pendulum, find the point where the tension of the string equals the weight of the particle.

15. *A particle is placed in a smooth tube which revolves horizontally about an axis through one end of it; required the equation of the curve traced by the particle.*

The only force to impel the particle along the tube is the centrifugal force due to rotation.

Letting r = the radius vector of the curve;

r_0 = the initial radius vector;

ω = the uniform angular velocity;

we have

$$\frac{d^2r}{dt^2} = \omega^2 r;$$

which, integrated, gives

$$r = \tfrac{1}{2} r_0 \left(e^{\omega t} + e^{-\omega t} \right),$$

hence, the relation between the radius vector and the arc described by the extremity of the initial radius vector, is the same as between the coördinates of a catenary. (See equation (k), p. 134.)

16. To find a curve joining two points down which a particle will slide by the force of gravity in the shortest time.

The curve is a cycloid. This problem is celebrated in the

history of Dynamics. The solution properly belongs to the Calculus of Variations, although solutions may be obtained by more elementary mathematics. Such curves are called *Brachistochrones*.

PROBLEMS PERTAINING TO THE EARTH.

122. *To find the value of g.*
We have, from example 10 of the preceding Article,

$$g = \frac{\pi^2 l}{T^2}. \qquad (149)$$

Making $T = 1$ second and $l = 39.1390$ inches, the length of the pendulum vibrating seconds at the Tower of London, we have for that place,

$$g = 32.1908 \text{ feet.}$$

The relation between the force of gravity at different places on the surface of the earth is given in Article 19.

The determination of l depends upon the compound pendulum.

123. *To find the centrifugal force at the equator.*
We have, from equation (142), for a unit of mass,

$$f = \omega^2 R = \frac{4\pi^2}{T^2} R; \qquad (a)$$

in which R, the equatorial radius, is $20{,}923{,}161$ feet; T, the time of the revolution of the earth on its axis, is $86{,}164$ seconds, and $\pi = 3.1415926$. These values give

$$f = 0.11126 \text{ feet.}$$

The force of gravity at the equator has been found to be 32.09022 feet (Article 19); hence, if it were not diminished by the centrifugal force, it would be

$$G = 32.09022 + 0.11126 = 32.20148 \text{ feet,}$$

and

$$\frac{f}{G} = \frac{0.11126}{32.20148} = \frac{1}{289} \text{ nearly};$$

hence the centrifugal force at the equator is $\frac{1}{289}$ of the undiminished force of gravity.

EXAMPLE.

In what time must the earth revolve that the centrifugal force at the equator may equal the force of gravity?

Ans. $\frac{1}{17}$ of its present time.

124. *To find the effect of the centrifugal force at different latitudes on the earth.*

Let $L = POQ =$ the latitude of the point P;

$R = OQ = OP =$ the radius of the earth;

then will the radius of the parallel of latitude PP' be

$R_1 = R \cos L$.

The centrifugal force will be in the plane of motion and may be represented by the line Pr, or

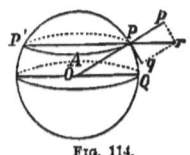

Fig. 114.

$$Pr = f_1 = \omega^2 R_1 = \omega^2 R \cos L;$$

therefore, the centrifugal force varies directly as the cosine of the latitude. But the force of gravity is in the direction PO. Resolving Pr parallel and perpendicular to PO, we have

$$Pp = \omega^2 R \cos^2 L = \tfrac{1}{289} G \cos^2 L;$$

$$Pq = \omega^2 R \cos L \sin L = \tfrac{1}{578} G \sin 2l;$$

the former of which diminishes directly the force of gravity, and the latter tends to move the matter in the parallel of latitude PP', toward the equator. Such a movement has taken place, and as a result the earth is an oblate spheroid. In the present form of the earth the action-line of the force of gravity is normal to the surface (or it would be if the earth were homogeneous), and hence, does not pass through the centre O, except on the equator and at the poles. The preceding formulas would be true for a rigid homogeneous sphere, but are only approximations in the case of the earth.

CHAPTER X.

FORCES IN A PLANE PRODUCING ROTATION.

125. ANGULAR MOTION OF A PARTICLE ABOUT A FIXED AXIS.
Let the body C, on the horizontal arm AB, revolve about the vertical axis ED. Consider the body reduced to the centre of the mass, and the force F_1 applied at the centre and acting continually tangent to the path described by the particle. This may be done as shown in Fig. 121. In this case the force will be measured in the same way as if the path were rectilinear, for the force is applied along the path. Hence, according to equation (21),

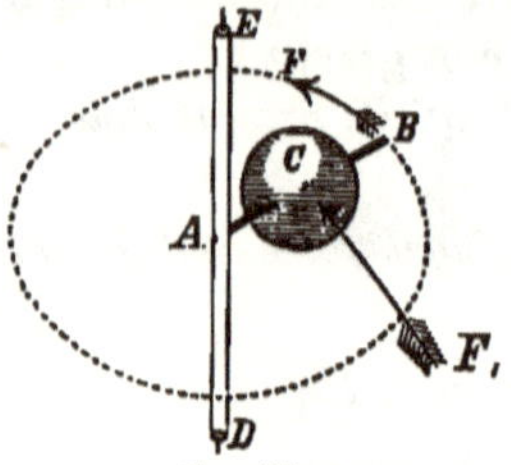

FIG. 115.

$$F_1 = m \frac{d^2 s}{dt^2};$$

in which s is the arc of the circle.

If θ be the angle swept over by the radius, and r_1 the radius; then

$$s = r_1 \theta,$$
$$ds = r_1 d\theta,$$
$$d^2 s = r_1 d^2 \theta;$$

which, substituted in the equation above, gives

$$F_1 = m r_1 \frac{d^2 \theta}{dt^2};$$

$$\therefore \frac{d^2 \theta}{dt^2} = \frac{F_1}{m r_1}. \tag{150}$$

If a force, F, be applied to the arm AB, at a distance a from the axis ED, producing the same movement of the mass C, we have, from the equality of moments, Article 65,

$$F_1 r_1 = Fa;$$

and the value of F_1 deduced from this equation substituted in the preceding one, gives

$$\frac{d^2\theta}{dt^2} = \frac{Fa}{mr_1^2}; \qquad (151)$$

that is, *the angular acceleration produced by a force, F, on a particle, m, equals the moment of the force divided by the moment of inertia of the mass.*

(For moments of inertia, see Chapter VII.)

We observe that, when the force is applied directly to the particle, it produces no strain upon the axis, but that it does when applied to other points of the arm. In both cases there will be a strain of

$$mr_1 \left(\frac{d\theta}{dt}\right)^2 = mr_1 \omega^2;$$

due to centrifugal force, ω being the angular velocity.

126. Angular motion of a finite mass. Let a body, AB, turn about a vertical axis at A, under the action of constant forces, F, acting horizontally; m_1, m_2, etc., masses of the elements of the body at the respective distances r_1, r_2, etc., from the axis A; and considering equation (151) as *typical*, we have

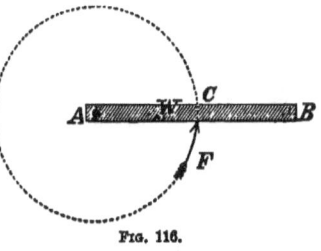

Fig. 116.

$$\frac{d^2\theta}{dt^2} = \frac{\Sigma Fa}{m_1 r_1^2 + m_2 r_2^2 + m_3 r_3^2 + \text{etc.}} = \frac{\Sigma Fa}{\Sigma mr^2}$$

$$= \frac{\text{moment of the forces}}{\text{moment of inertia}}. \qquad (152)$$

127. Energy of a rotating mass. Multiply both members

of the preceding equation by $d\theta$, integrate and reduce, and we find

$$\tfrac{1}{2}\Sigma mr^2 \left(\frac{d\theta}{dt}\right)^2 = \int F \cdot r d\theta \; ; \qquad (153)$$

in which $rd\theta$ is an element of the space passed over by F, and $F \cdot rd\theta$ is an element of work done by F; hence the second member represents the total work done by F upon the body, all of which is stored in it. Therefore, *the energy of a body rotating about an axis equals the moment of inertia of the mass multiplied by one-half the square of the angular velocity.*

If the body has a motion of translation and of rotation at the same time, the total energy will be the sum due to both motions; for it is evident that while a body is rotating a force may be applied to move it forward in space, Article 38, and that the work done by this force will be independent of the rotation.

If $v =$ the velocity of translation of the axis about which the body rotates;

$\omega =$ the angular velocity; and

$I_m =$ the moment of inertia of the *mass*;

then the total work stored in the body will be

$$\tfrac{1}{2}Mv^2 + \tfrac{1}{2}I_m\omega^2 \qquad (154)$$

128. An impulse. Multiply both members of equation (152) by dt, integrate, and we find

$$\frac{d\theta}{dt} = \omega = \frac{\Sigma a \int F dt}{\Sigma mr^2}.$$

But according to Article 33, $\int F dt$ is the measure of an impulse and is represented by Q, hence

$$\omega = \frac{Qa}{\Sigma mr^2} \qquad (155)$$

$$= \frac{\textit{moment of the impulse}}{\textit{moment of inertia}}$$

129. The time required to pass over n circumferences will be, in the case of an impulse,

$$t = \frac{2n\pi}{\omega} = \frac{2n\pi I}{Qa}.$$

In the case of accelerated forces the time will be found by integrating equation (153).

130. SIMULTANEOUS MOVEMENT OF ROTATION AND OF TRANSLATION. If the body be unconstrained, the motion of translation of the body will be that of the centre of the mass. If the axis of rotation is rigid, it may be located anywhere in the body, or even without the body by considering it as rigidly connected with the body, in which case the motion of translation will be that of some point of the axis. In either case the motion of translation may be considered as resulting from a force acting directly upon the axis of rotation, and the rotation, by a force acting at some other point. The two motions may then be considered as existing independently of each other.

131. FORMULAS for the movement of a body involving both translation and rotation. The *general* equations for this case are (164) and (165) in the next chapter. In this Article let the rotation be about an axis parallel to z, and the centre of the mass move in the plane xy. Resolve the forces into couples and forces applied at the origin of coördinates, as in Article 83; then will the third of equations (86) be the impressed forces which produce rotation. Let R be the resultant of the forces at the origin at any instant, (see Article 84), and s be measured along the path described by the point of intersection of the axis of rotation with the plane xy. Then will the principles of Article 21, give

$$\left. \begin{array}{r} R - \Sigma m \dfrac{d^2 s}{dt^2} = 0; \\[6pt] \Sigma (Xy - Yx) - \Sigma \left(my \dfrac{d^2 x}{dt^2} - mx \dfrac{d^2 y}{dt^2} \right) = 0; \end{array} \right\} \quad (150)$$

The expression $\Sigma (Xy - Yx)$ is the sum of the moments of the impressed forces $= \Sigma Fa$ (Article 60). Transforming the second term of the last equation into polar coördinates having the same origin, we have

$$\Sigma \left(my \frac{d^2 x}{dt^2} - mx \frac{d^2 y}{dt^2} \right) = \Sigma m r^2 \frac{d^2 \theta}{dt^2};$$

hence, the equations become

$$m\frac{d^2s}{dt^2} = F;$$
$$\frac{d^2\theta}{dt^2} = \frac{\Sigma Fa}{\Sigma mr^2}.$$
(157)

132. REDUCED MASS. A given mass may be concentrated at such a point, or in a thin annulus, that the force or impulse will have the same effect upon it as if it were distributed. To accomplish this it is only necessary that Σmr^2 in the second of (157) should have an equivalent value. Let M be the mass of the body, k the distance from the axis to the required point, then

$$\Sigma mr^2 = Mk^2,$$

in which k is the radius of gyration, as defined in Article 107.

But any other point may be assumed, and a *mass determined* such that the effect shall be the same. Let κ be the distance to that point (or radius of the annulus), and M_1 the required mass, then we have

$$Mk^2 = M_1\kappa^2;$$
$$\therefore M_1 = M\frac{k^2}{\kappa^2};$$
(158)

which is called the reduced mass.

EXAMPLES.

1. *A prismatic bar AB, falls through a height h, retaining its horizontal position until one end strikes a fixed obstacle C; required the angular velocity of the piece and the linear velocity of the centre immediately after the impulse.*

Let M be the mass of the bar, l its length, v the velocity of the centre at the instant of impact, and v_1 the velocity of the centre immediately after impact. Consider the bodies as perfectly non-elastic; then will the effect of the impact be simply to suddenly arrest the end A.

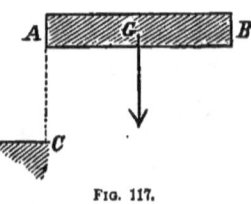

FIG. 117.

The bar will rotate about a horizontal axis through the centre, as shown by Article 38; and, as shown by Articles 27 and 38, the impulse will be $Q = M(v - v_1)$; that is, *it is the change of velocity at the centre multiplied by the mass.* The impact will entirely arrest the motion of the end, A, at the *instant* of the impact, and hence at that instant the angular velocity of A in reference to G will be the same as G in reference to A.

Equation (155) gives

$$\omega = \frac{\text{moment of impulse}}{\text{moment of inertia}}$$

$$= \frac{M(v - v_1) \tfrac{1}{2}l}{\tfrac{1}{12}Ml^2}$$

$$= 6\frac{v - v_1}{l}.$$

But at the instant of the impact

$$v_1 = \tfrac{1}{2}l\omega,$$

solving these give

$$\omega = \tfrac{3}{2}\frac{v}{l}, \quad v_1 = \tfrac{3}{4}v.$$

We now readily find

$$Q = \tfrac{1}{4}Mv.$$

To find the velocity of any point in a vertical direction at the instant of the impact, we observe that it may be considered as composed of two parts; a linear velocity v_1 downward, and a right-handed rotation. The actual velocity at A due to rotation will be

$$\tfrac{1}{2}l\omega = \tfrac{3}{4}v,$$

which will be upward, and the linear velocity downward will be $v_1 = \tfrac{3}{4}v$, hence the result will be no velocity. Similarly, the velocity at B will be $\tfrac{3}{4}v + \tfrac{3}{4}v = \tfrac{3}{2}v$. Also, for any point distant x from G, we have at the left of G

$$\tfrac{3}{4}v - \omega x = \tfrac{3}{4}v\left(1 - 2\frac{x}{l}\right);$$

and to the right of G we have

$$\tfrac{3}{4}v\left(1 + 2\frac{x}{l}\right).$$

When the bar comes into a vertical position, we easily find

that A has passed below a horizontal through C. Every point, therefore, has a progressive velocity, except the point A, at the instant of impact.

After the impact the centre will move in the same vertical and with an accelerated velocity, while the angular velocity will remain constant.

2. Suppose that impact takes place at one-quarter the length from A, required the angular velocity.

3. At what point must the impulse be made so that the velocity of the extremity B will be doubled at the instant of impact?

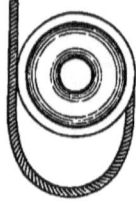

Fig. 118.

4. An inextensible string is wound around a cylinder, and has its free end attached to a fixed point. The cylinder falls through a certain height (not exceeding the length of the free part of the string), and at the instant of the impact the cord is vertical and tangent to the cylinder; all the forces being in a plane; required the angular velocity produced by the impulse, and the momentum.

$$Ans. \ \tfrac{3}{8}\frac{v}{r}; \quad Q = \tfrac{1}{8}Mv.$$

5. In the preceding problem, let the body be a homogeneous sphere, the string being wound around the arc of a great circle.

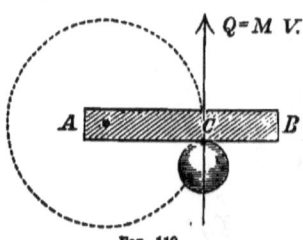

Fig. 119.

6. *A homogeneous prismatic bar AB, in a horizontal position constrained to revolve about a vertical fixed axis A, receives a direct impulse from a sphere whose momentum is Mv; required the angular velocity of the bar.*

The momentum imparted to the bar will depend upon the elasticities of the two bodies. Consider them perfectly elastic. The effect of the impulse will be the same as if the mass of the bar were concentrated at the extremity of the radius of gyration; hence an equivalent mass at the point C may be determined.

Let M_1 = the mass of AB;
M_2 = the reduced mass;
v_2 = the velocity of the reduced mass after impact;
$a = AC$.

Then, by equation (158), the mass of the bar reduced to the point C, will be

$$M_2 = M_1 \frac{k^2}{a^2}.$$

By equation (40) the velocity of M_2 after impact will be

$$v_2 = \frac{2M}{M + M_2} v = \frac{2Ma^2}{Ma^2 + M_1 k^2} v; \quad (a)$$

hence, the momentum imparted will be

$$M_2 v_2 = \frac{2MM_1 k^2}{Ma^2 + M_1 k^2} v;$$

and the moment will be

$$M_2 v_2 a = \frac{2MM_1 k^2 a}{Ma^2 + M_1 k^2} v.$$

According to equation (155), we have

$$\omega = \frac{\text{moment of the impulse}}{\text{moment of inertia}}$$

$$= \frac{\frac{2MM_1 k^2 a}{Ma^2 + M_1 k^2} v}{M_1 k^2}$$

$$= \frac{2Ma}{Ma^2 + M_1 k^2} v.$$

This result is the same as that found by dividing equation (a) by a, as it should be.

7. *Suppose, in the preceding problem, that there is no fixed axis, but that the body is free to translate; find where the impact must be made that the initial velocity at the end A shall be zero.*

Let Mv be the impulse *imparted* to the body;
$Mk_1^2 =$ the principal moment of inertia;
$h =$ the distance from the centre of the bar to the required point;
then

$$\omega = \frac{\text{moment of the impulse}}{\text{moment of inertia}}$$

$$= \frac{Mvh}{Mk_1^2} = \frac{vh}{k_1^2}; \qquad (a)$$

and the movement at A in the circular arc will be

$$\tfrac{1}{2}l\omega = \frac{vlh}{2k_1^2};$$

and the initial linear movement will be

$$v - \frac{vlh}{2k_1^2};$$

which, by the conditions of the problem, will be zero; hence

$$\frac{vlh}{2k_1^2} = v$$

or,

$$h = \frac{2k_1^2}{l}. \qquad (b)$$

The distance from A will be

$$h + \tfrac{1}{2}l = \frac{k_1^2 + (\tfrac{1}{2}l)^2}{\tfrac{1}{2}l}.$$

The bar being prismatic, $k_1^2 = \tfrac{1}{12}l^2$;

$$\therefore h + \tfrac{1}{2}l = \tfrac{2}{3}l.$$

The result is independent of the magnitude of the impulse. From (b) we have

$$h(\tfrac{1}{2}l) = k_1^2;$$

hence, h and $\tfrac{1}{2}l$ are convertible, and we infer that if the impulse be applied at A the point of no initial motion will be at the point given by equation (*b*), where the impact was previously applied.

8. *In the preceding problem find where the impulse must be applied so that the point of no initial velocity shall be at a distance h' from the centre.*

The initial linear velocity due to the rotary movement found from (*a*) of the preceding example, will be

$$h'\omega = \frac{v}{k_1^2} hh',$$

and the initial movement of the required point being zero, we have

$$v - \frac{v}{k_1^2} hh' = 0;$$

$$\therefore h = \frac{k_1^2}{h'}. \qquad (159)$$

If the point of impact be at b, the point a, where the initial movement is zero, will be on the other side of the centre of the body. Let B be the centre, then

$$h = bB, \qquad h' = aB,$$

and from (159), we have

$$hh' = bB.aB = k_1^2; \qquad (160)$$

and as k_1 is a constant, *the points a and b are convertible.*

AXIS OF SPONTANEOUS ROTATION.

133. In the preceding problem the *initial* motion would have been precisely the same if there had been a fixed axis through *a* perpendicular to the plane of motion, and hence the *initial motion* may be considered as a rotation about that axis. If a fixed axis were there it evidently would not receive any shock from the impulse.

The axis about which a quiescent body tends to turn at the instant that it receives an impulse is called the axis of spontaneous rotation.

CENTRE OF PERCUSSION.

134. When there is a fixed axis and the body is so struck that there is no impulse on the axis, *any point in the action-line of the force is called the centre of percussion.* Thus in Fig. 120, if a is the fixed axis, b will be the centre of percussion. It is also evident that, if b be a fixed object, and it be struck by the body AC, rotating about a, the axis will not receive an impulse.

AXIS OF INSTANTANEOUS ROTATION.

135. An axis through the centre of the mass, parallel to the axis of spontaneous rotation, is called the *axis of instantaneous rotation.* A free body rotates about this axis.

In regard to the spontaneous axis, we consider *that* as fixed in space for the instant; but at the same time the body really rotates about the instantaneous axis which moves in space with the body.

EXAMPLES UNDER THE PRECEDING EQUATIONS CONTINUED.

9. *A horizontal uniform disc is free to revolve about a vertical axis through its centre. A man walks around on the outer edge; required the angular distance passed over by the man and disc when he has walked once around the circumference.*

Let $W =$ the weight of the man;
$w =$ the weight of the disc;
$r =$ the radius of the disc;
$\omega_1 =$ the angular velocity of the man in reference to a fixed line;
$\omega =$ the angular velocity of the disc in reference to the same fixed line;
$Q =$ the force exerted by the man against the disc;
$k_1^2 = \tfrac{1}{2} r^2$.

The result will be the same whether the effort be exerted suddenly, or with a uniform acceleration, or irregularly. We will, therefore, treat it as if it were an impulse. The weights are here used instead of the masses, for they are directly proportional to each other, and it is more natural to speak of the weight of a man than the mass of a man.

We have

$$\omega = \frac{\text{moment of the impulse}}{\text{moment of inertia}}$$

$$= \frac{Qr}{wk_1^2}$$

$$= \frac{Wvr}{wk_1^2}$$

$$= \frac{2W}{w}\omega_1.$$

But when the man arrives at the initial point of the disc, we have

$$\omega + \omega_1 = 2\pi;$$

which, combined with the preceding equation, gives

$$\omega_1 = \frac{2w\pi}{w + 2W}.$$

If $W = w$, we have

$$\omega_1 = \tfrac{4}{3}\pi,$$

for the angular space passed over by the man, and

$$\omega = \tfrac{2}{3}\pi,$$

for the distance passed over by the disc.

10. *In Fig. 115 let the force F be constant; required the number of complete turns which the body C will make about the axis DE in the time t.*

Let $r =$ the radius of the circle passed over by F;
$\quad r_1 =$ the distance of the centre of the body from the axis of revolution;
$\quad k_1 =$ the principal radius of gyration of the body in reference to a moment axis parallel to DE;
$\quad k =$ the radius of gyration of the body in reference to the axis DE;
then, according to equation (123),

$$k^2 = r_1^2 + k_1^2;$$

and, according to equation (152),

$$\frac{d^2\theta}{dt^2} = \frac{\text{moment of forces}}{\text{moment of inertia}}$$

$$= \frac{Fr}{Mk^2}.$$

Multiply by dt and integrate, and we have

$$\frac{d\theta}{dt} = \frac{Fr}{Mk^2}t.$$

the constant being zero, for the initial quantities are zero.

Multiplying again by dt, we find

$$\theta = \frac{Fr}{2Mk^2}t^2;$$

which is the angular space passed over in time t; and the number of complete rotations will be

$$\frac{\theta}{2\pi} = \frac{Frt^2}{4\pi Mk^2}.$$

11. If the body were a sphere 2 feet in diameter, weighing 100 pounds, the centre of which was 5 feet from the axis; F, a force of 25 pounds, acting at the end of a lever 8 feet long; required the number of turns which it will make about the axis in 5 minutes.

12. If the data be the same as in the preceding example; required the time necessary to make one complete turn about the axis.

13. Suppose that an indefinitely thin body, whose weight is W, rests upon the rim of a horizontal pulley which is perfectly free to move. A string is wound around the pulley, and passes over another pulley and has a weight, P, attached to its lower end. Supposing that there is no resistance by the pulleys or the string, required the distance passed over by P in time t.

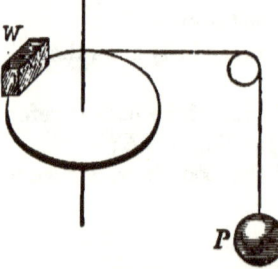

Fig. 121

[135.] SOLID BODIES. 213

According to equation (152), we have

$$\frac{d^2\theta}{dt^2} = \frac{Pr}{\frac{W+P}{g}r^2} = \frac{Pg}{(W+P)r};$$

from which it may be solved.

(This is equivalent to applying the weight P directly to the weight W, as in Fig. 10, and hence we have, according to equation (21),

$$\frac{W+P}{g}\frac{d^2s}{dt^2} = P;$$

but referring it to polar coördinates, we have $r\frac{d^2\theta}{dt^2} = \frac{d^2s}{dt^2}$, which substituted reduces the equation directly to that in the text.)

14. A disc whose weight is W is free to revolve about a horizontal axis passing through its centre and perpendicular to its plane. A cord is wound around its circumference and has a weight, P, attached to its lower end; required the distance through which P will descend in t seconds.

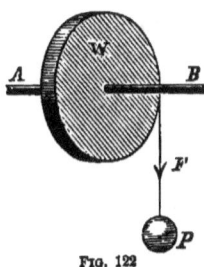

Fig. 122

We have

$$\frac{d^2\theta}{dt^2} = \frac{Prg}{Wk_1^2 + Pr^2};$$

from which θ may be found, and the space will be $r\theta$.

(This may be solved by equation (21). The mass of the disc reduced to an equivalent at the circumference will be $\dfrac{W}{g}\dfrac{k_1^2}{r^2}$, and that equation will become

$$\frac{1}{g}\left(P + W\frac{k_1^2}{r^2}\right)\frac{d^2s}{dt^2} = P;$$ which, by changing to polar coördinates, may be reduced to the equation in the text.)

15. If, in the preceding example, the body were a sphere revolving about a horizontal axis, the diameter of the sphere being 16 inches, weight 500 pounds, moved by a weight of 100 pounds descending vertically, the cord passing around a groove in the sphere the diameter of which is one foot; required the number of revolutions of the sphere in five seconds.

16. 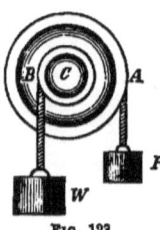 Two weights, P and W, are suspended on two pulleys by means of cords, as shown in Fig. 123, the pulleys being attached to the same axis C. No resistance being allowed for the pulleys, axle, or cords; required the circumstances of motion.

Fig. 123.

We have

$$\frac{d^2\theta}{dt^2} = \frac{\text{moment of the forces}}{\text{moments of inertia}}$$

$$= \frac{P.AC - W.BC}{P(AC)^2 + W(BC)^2 + \text{disc } AC.k_1^2 + \text{disc } BC.k_2^2} g \, ;$$

in which *disc AC*, etc., are used for the weights of the discs. Let the right-hand member be represented by M, then we have

$$\frac{d\theta}{dt} = \omega = Mt;$$

$$\theta = \tfrac{1}{2}Mt^2.$$

17. In the preceding example let the discs be of uniform density, $AC = 8$ inches, $BC = 3$ inches; the weight of $AC = 6$ pounds, of $CB = 2$ pounds, of $P = 25$ pounds, and of $W = 60$ pounds; if they start from rest, required the space passed over by P in 10 seconds, and the tension of the cords.

18. *A homogeneous, hollow cylinder rolls down an inclined plane by the force of gravity; required the time.*

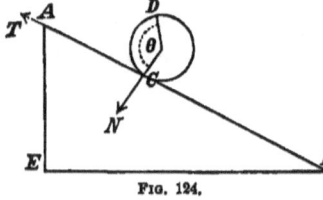

Fig. 124.

The weight of the cylinder may be resolved into two components, one parallel to the plane, which impels the body down it, the other normal, which induces friction. The friction acts parallel to the plane and tends to prevent the movement down it, and is assumed to be sufficient to prevent sliding.

Let $W =$ the weight of the cylinder;
    $i =$ the inclination of the plane to the horizontal;
    $N = W \cos i =$ the normal component;
    $m =$ the mass of a unit; the altitude $= 1$;

[135.]   SOLID BODIES.   215

$\phi =$ the coefficient of friction;
$T = \phi N =$ the tangential component;
$T' = W \sin i =$ the component of the weight parallel to the plane;
$r_1 =$ the internal radius of the cylinder;
$r =$ the external radius;
$\theta =$ the angular space passed over by the radius;
$s = AC$, the space.

This is a case of translation and rotation combined, and equations (157) give

$$\frac{W}{g}\frac{d^2s}{dt^2} = T' - T = W \sin i - T;$$

$$\frac{d^2\theta}{dt^2} = \frac{T.r}{\tfrac{1}{2}m\pi (r^4 - r_1^4)} = \frac{2gTr}{W(r^2 + r_1^2)};$$

and from the problem

$$s = r\theta.$$

Eliminating s and T from these equations, we get

$$\frac{d^2\theta}{dt^2} = \frac{2gr \sin i}{3r^2 + r_1^2}.$$

Integrating and making the initial spaces zero, we have

$$\theta = \frac{gr \sin i}{3r^2 + r_1^2} t^2;$$

$$\therefore t = \sqrt{\frac{3r^2 + r_1^2}{gr^2 \sin i}}\, s.$$

If $r_1 = 0$, the cylinder will be solid, and

$$t = \sqrt{\frac{3s}{g \sin i}},$$

and hence, the time is independent of the diameter of the cylinder.

If $r_1 = r$, the cylinder will be a thin annulus, and

$$t = \sqrt{\frac{4s}{g \sin i}};$$

hence, the time of descent will be $\sqrt{\tfrac{4}{3}}$ times as long as when the cylinder is solid; the weight being the same in both cases.

If it slide down a smooth plane of the same slope, we have

$$t = \sqrt{\frac{2s}{g \sin i}},$$

which is less than either of the two preceding times.

THE PENDULUM.

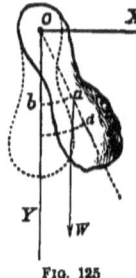

Fig. 125

18. *Let a body be suspended on a horizontal axis and moved by the force of gravity; required the circumstances of motion.*
We have

$$-\frac{d^2\theta}{dt^2} = \frac{\text{moment of forces}}{\text{moment of inertia}}$$

$$= \frac{Wh \sin \theta}{Mk^2};$$

in which

$h = Oa$, the distance from the axis of suspension to the centre of gravity a of the body;
$W =$ the weight of the body;
$\theta = bOa$; and let
$k_1 =$ the principal radius of gyration;

then the preceding equation becomes

$$-\frac{d^2\theta}{dt^2} = \frac{gh}{h^2 + k_1^2} \sin \theta.$$

This equation cannot be completely integrated in finite terms, but by developing $\sin \theta$ and neglecting all powers above the first, we find for a complete oscillation

$$T = \pi \sqrt{\frac{h^2 + k_1^2}{gh}}; \qquad (161)$$

which gives the time in seconds when h, k_1 and g are given in feet.

To find the length of a simple pendulum which will vibrate

in the same time, we make equations (b), page 196, and (161) equal to one another, and have

$$l = \frac{h^2 + k_1^2}{h} = Od. \quad (162)$$

Let $ad = h_1$, then

$$l - h = h_1 = \frac{k_1^2}{h} = ag,$$

$$\therefore hh_1 = k_1^2 \quad (163)$$

136. DEFINITIONS. A body of any form oscillating about a fixed axis is called a *compound pendulum*.

A material particle suspended from a string without weight, oscillating about a fixed axis, is called a *simple pendulum*.

The point d is called the *centre of oscillation*. It is the point at which, if a particle be placed and suspended from the axis O by a string without weight, it will oscillate in the same time as the body Od. Or, it is the point at which, if the entire mass be concentrated, it will oscillate about the axis in the same time as when it is distributed.

The point O, where the axis pierces the plane xy, is called the *centre of suspension*.

137. RESULTS. The centres of oscillation and of percussion coincide. (See Article 134.)

According to equation (163), the centres of oscillation and of suspension are convertible.

According to the same equation the principal radius of gyration is a mean proportional between the distances of oscillation and of suspension from the centre of gravity.

Equation (161) indicates a practical mode of determining the principal radius of gyration. To find it, let the body oscillate, and thus find T, then attach a pair of spring balances to the lower end and bring the body to a horizontal position, and find how much the scales indicate; knowing which, the weight of the body and the distance between the point of attachment and the centre of suspension O, the value of h may easily be computed. The value of g being known, all the quantities in equation (161) become known except k_1, which is readily found by a solution of the equation.

EXAMPLES.

1. A prismatic bar oscillates about an axis passing through one end, and perpendicular to its length; required the length of an equivalent simple pendulum.

2. A homogeneous sphere is suspended from a point by means of a fine thread, find the length of a simple pendulum which will oscillate in the same time.

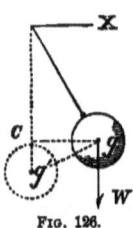

FIG. 126.

138. Captain Kater used the principle of the convertibility of the centres of suspension and oscillation for determining the length of a simple seconds pendulum, and hence the acceleration due to gravity.—*Phil. Trans.*, 1818.

Let a body, furnished with a movable weight, be provided with a point of suspension C (figure not shown), and another point on which it may vibrate, fixed as nearly as can be estimated in the centre of oscillation O, and in a line with the point of suspension and the centre of gravity. The oscillations of the body must now be observed when suspended from C and also when suspended from O. If the vibrations in each position should not be equal in equal times, they may readily be made so by shifting the movable weight. When this is done, the distance between the two points C and O is the length of the simple equivalent pendulum. Thus the length L and the corresponding time T of vibration will be found uninfluenced by any irregularity of density or figure. In these experiments, after numerous trials of spheres, etc., knife edges were preferred as a means of support. At the centres of suspension and oscillation there were two triangular apertures to admit the knife edges on which the body rested while making its oscillations.

Having thus the means of measuring the length L with accuracy, it remains to determine the time T. This is effected by comparing the vibrations of the body with those of a clock. The time of a single vibration or of any small arbitrary number of vibrations cannot be observed directly, because this would require the fraction of a second of time, as shown by the clock, to be estimated either by the eye or ear. The vibrations of the body may be counted, and here there is no fraction to be estimated, but these vibrations will not probably fit in with the oscillations of the clock pendulum, and the differences must be estimated. This defect is overcome by "the method of coincidences." Supposing the time of vibration of the clock to be a little less than that of the body, the pendulum of the clock will gain on the body, and at length at a certain vibration the two will for an instant coincide. The two pendulums will now be seen to separate, and after a time will again approach each other, when the same phenomenon will take place. If the two pendulums continue to vibrate with perfect uniformity, the number of oscillations of the pendulum of the clock in this interval will be an integer, and the number of oscillations of the body in the same interval will be less by one complete oscillation than that of the pendulum of the clock. Hence by a simple proportion the time of a complete oscillation may be found.

The coincidences were determined in the following manner: Certain marks made on the two pendulums were observed by a telescope at the lowest point of their arcs of vibration. The field of view was limited by a diaphragm to a narrow aperture across which the marks were seen to pass. At each succeeding vibration the clock pendulum follows the other more closely, and at last the clock-mark completely covers the other during their passage across the field of view of the telescope. After a few vibrations it appears again preceding the other. The time of disappearance was generally considered as the time of coincidence of the vibrations, though in strictness the mean of the times of disappearance and reappearance ought to have been taken, but the error thus produced is very small. (*Encyc. Met.* Figure of the Earth.) In the experiments made in Hartan coal-pit in 1854, the Astronomer Royal used Kater's method of observing the pendulum. (*Phil. Trans.*, 1856.)

The value of T thus found will require several corrections. These are called "Reductions." If the centre of oscillation does not describe a cycloid, allowance must be made for the alteration of time depending on the arc described. This is called "the reduction to infinitely small arcs." If the point of support be not absolutely fixed, another correction is required

(*Phil. Trans.*, 1831). The effect of the buoyancy and the resistance of the air must also be allowed for. This is the "reduction to a vacuum." The length L must also be corrected for changes of temperature.

The time of an oscillation thus corrected enables us to find the value of gravity at the place of observation. A correction is now required to reduce this result to what it would have been at the level of the sea. The attraction of the intervening land must be allowed for by Dr. Young's rule (*Phil. Trans.*, 1819). We thus obtain the force of gravity at the level of the sea, supposing all the land above this level were cut off and the sea constrained to keep its present level. As the sea would tend in such a case to change its level, further corrections are still necessary if we wish to reduce the result to the surface of that spheroid which most nearly represents the earth. (See *Camb. Phil. Trans.*, vol. x.)

There is another use to which the experimental determination of the length of a simple equivalent pendulum may be applied. It has been adopted as a standard of length on account of being invariable and capable at any time of recovery. An Act of Parliament, 5 Geo. IV., defines the yard to contain thirty-six such parts, of which parts there are 39.1393 in the length of the pendulum vibrating seconds of mean time in the latitude of London, in vacuo, at the level of the sea, at temperature 62° F. The Commissioners, however, appointed to consider the mode of restoring the standards of weight and measure which were lost by fire in 1834, report that several elements of reduction of pendulum experiments are yet doubtful or erroneous, so that the results of a convertible pendulum are not so trustworthy as to serve for supplying a standard for length; and they recommend a material standard, the distance, namely, between two marks on a certain bar of metal under given circumstances, in preference to any standard derived from measuring phenomena in nature. (*Report*, 1841.)

All nations, practically, use this simple mode of determining the length of the standard of measure, that of placing two marks on a bar, and by a legal enactment declaring it to be a certain length.

139. FORM OF THE EARTH. The pendulum furnishes one of the best means for determining the form of the earth.

Let a = the equatorial radius of the earth;
b = the semi-axis;
ϵ = the ellipticity of the earth;
then
$$\epsilon = \frac{a-b}{a}.$$

Let m = ratio of the centrifugal force at the equator to the force of gravity at the same place;
l_0 = length of a second's pendulum at the equator;
l_{90} = the length of a second's pendulum at the poles;
then, from the *Mécanique Céleste*, tome II., No. 34, we have
$$\epsilon = \tfrac{5}{2} m - \frac{l_{90} - l_0}{l_0}.$$

The value of m is $\frac{1}{289}$. The formula for the length of the second's pendulum when the length at Paris is taken as unity, is
$$l = 0.996823 + 0.00549745 \sin^2 H,$$
when H is the latitude of the place. See Puissant's *Traité de Géodésie*, page 461.

By this means it has been found that ϵ is about $\frac{1}{317}$. Bowdich, in his translation of the *Mécanique Céleste*, p. 485, remarks, " It appears that the oblateness (ϵ) does not differ much from $\frac{1}{300}$, and may possibly be a little more, though some results give a little less."

140. TORSION PENDULUM. If an elastic bar, CD, be fixed at one end, and at the other end have two weights, A_1 and A_2, rigidly fixed to it by means of the cross arm, $A_1 A_2$, then if the arm be turned into the position $B_1 B_2$, the elastic resistance of the bar DC will cause the weights to move back to $A_1 A_2$, and by virtue of the energy of the weights at that point, they will pass that position, and move on until their motion is arrested by the action of the elastic resistance of the

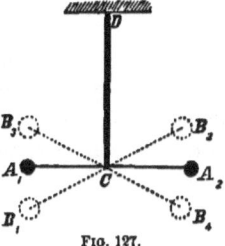

FIG. 127.

bar; after which they will return to their former position, thus having a motion similar to that of the common pendulum. This arrangement is called a *torsion pendulum*. The motion will be the same for one weight as for two, but when the bar DC is vertical, the arms CA_1 and CA_2 should equal each other, and the weight A_1 equal A_2.

141. *To find the force necessary to twist the rod DC through a given angle.*

Let $F=$ the force at A_1 perpendicular to the arm CA_1;
$a = CA_2 = CA_1$; $l = DC$;
$a = A_1CB_1$; $I =$ the polar moment of inertia of a transverse section of the bar DC; and
$G =$ the coefficient of the elastic resistance to torsion.

The moment of the twisting force, F, will be

$$Fa;$$

and the moment of the elastic resistance will be (see *Resistance of Materials*, 2d Ed., p. 206),

$$GI\frac{a}{l};$$

hence, we have

$$Fa = GI\frac{a}{l};$$

$$\therefore F = GI\frac{a}{al}.$$

The weights A_1 and A_2 are not involved in this problem.

If the angle be measured from some fixed line making an angle ϕ with the neutral position of A_1A_2, then instead of a we would have $a_1 - \phi$, and the last equation becomes

$$\frac{Fa}{I} = \frac{G}{l}(a_1 - \phi).$$

If the force be reversed, it will twist the bar in the opposite direction, making an angle a_2 with the fixed line of reference, and we would have

$$\frac{Fa}{I} = \frac{G}{l}(\phi - a_2).$$

Adding these equations gives

$$2\frac{Fa}{I} = \frac{G}{l}(a_1 - a_2).$$

142. *To find the time of an oscillation.*

The bar CD having been twisted by moving the bar from its normal position A_1A_2 into the position B_1B_2, and then left to itself, it is required to find the time of moving to the other extreme position B_3B_4. We will neglect the mass of the rod A_1A_2, and that of the bar DC, and thus simplify the solution, and secure an approximate result.

Let $I_A = $ the moment of inertia of *one* of the bodies, A_1, or A_2, in reference to CD as an axis;

$\theta = $ a variable angle measured from the neutral position, A_1A_2; and,

considering Fa as a variable moment, producing the variable angle θ, we have from the second of equations (157), and the value of Fa from the first equation of the preceding Article,

$$-2I_A\frac{d^2\theta}{dt^2} = GI\frac{\theta}{l}.$$

Multiply by $d\theta$ and integrate, and observing that for $\theta = a$ the angular velocity is zero, we find

$$\frac{d\theta^2}{dt^2} = \frac{GI}{2I_Al}(a^2 - \theta^2);$$

hence

$$dt = \sqrt{\frac{2I_Al}{GI}}\frac{d\theta}{\sqrt{a^2-\theta^2}}.$$

Integrating again gives

$$t = \sqrt{\frac{2I_Al}{GI}}\sin^{-1}\frac{\theta}{a},$$

which, between the limits of 0 and a, gives

$$t = \tfrac{1}{2}\pi\sqrt{\frac{2I_A l}{GI}};$$

which is the time of half an oscillation; hence the time for a full oscillation, or the time of movement from B_1 to B_3, will be

$$2t = T = \pi\sqrt{\frac{2I_A l}{GI}};$$

which is the time required. The times are isochronous and independent of the amplitude.

The value of G may be eliminated by substituting its value taken from the last equation of the preceding article. Making the substitution, we find

$$T = \pi\sqrt{\frac{I_A(a_1 - a_2)}{Fa}};$$

from which l and I have also disappeared. From this we find

$$Fa = I_A(a_1 - a_2)\left(\frac{\pi}{T}\right)^2$$

143. *To find the density of the earth.*
The plan of determining the density of the earth by means of a torsion rod was first suggested by the Rev. John Mitchel. He died before he was able to make the experiment, but the plan was executed by Mr. Cavendish, who published the result in the *Phil. Trans.* for 1798. Subsequent to 1837, Mr. Bailey, at the request of the Astronomical Society (England), made a new determination of the result. He made upwards of 2,000 experiments with balls of different weights and sizes, and suspended in a variety of ways, a full account of which is given in the *Memoirs of the Astronomical Society*, Vol. xiv. We give here only some of the more prominent features of the experiment.

The torsion rod DC was very small, so that it could be easily twisted. Two small balls, A_1, A_2, were suspended from the torsion rod by a light cross-bar. Two large balls, E_1 E_2, were placed on a plank which turned about a point O directly under C, and the whole so arranged that the centres of gravity of the four balls were in the same horizontal plane. The apparatus

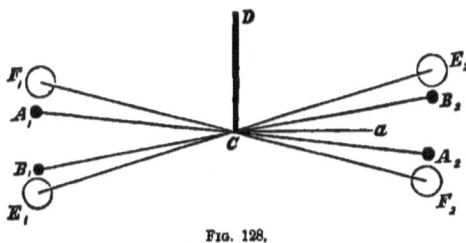

Fig. 128.

was inclosed in a small room so as to exclude currents of air, and the weights E_1 and E_2 were moved into the desired positions from the outside of the room by means of mechanism extending into the room.

The weights E_1 and E_2 were first placed nearly at right angles with the rod $A_1 A_2$, when the latter would assume some neutral position as Ca. The balls $E_1 E_2$ were then brought quite near to the small ones, B_1, B_2, when the attraction of the former drew the latter from their neutral position, and they oscillated about some position of equilibrium as $B_1 B_2$. The angle aCB_2 was observed, and also the time of the oscillation about the position CB_2.

The balls were then changed to the position $F_1 F_2$, making the angle aCF_2 as nearly equal as possible to aCE_2; but it was found that the line Ca did not always bisect the angle $E_2 CF_2$, but the mean of many readings was taken as the most probable value. The angle $E_2 CF_2$ will be $a_1 - a_2$, given in the preceding Article.

It is proved, by the law of attraction, that the attraction of a homogeneous sphere is the same as if its entire mass were concentrated at the centre of the mass, and varies inversely as the square of the distance from the centre.

Let $M =$ the mass of one of the large balls;
  $m =$ "   "   "   "   "   " small balls;

D = the distance between their centres; and
μ = the attraction of a sphere whose mass is unity upon another unit when the distance between their centres is unity;

then the force of attraction of the mass M upon m will be

$$\mu \frac{Mm}{D^2};$$

and this is the value of F in the last equation of the preceding Article. But there being two balls in this case, the *moment* of this attractive force will be

$$2\mu \frac{Mm}{D^2} a;$$

which (by neglecting the attraction of the large ball and plank upon the rod CB_2, and of the plank upon the small balls), equals the second member of the last equation of the preceding Article. Hence,

$$2\mu \frac{Mm}{D^2} a = I_{\text{A}}(a_1 - a_2)\left(\frac{\pi}{T}\right)^2$$

Let E be the mass of the earth, R its radius, and g the force of gravity, then

$$g = \mu \frac{E}{R^2}.^*$$

Eliminating μ, and making $I_{\text{A}} = m(a^2 + \tfrac{2}{5}r^2)$, we have

$$\frac{M}{E} = (a^2 + \tfrac{2}{5}r^2)(a_1 - a_2)\frac{D^2}{2R^2 ga}\left(\frac{\pi}{T}\right)^2$$

The density of the earth is thus reduced to the determination of the $a_1 - a_2$ between its two positions of equilibrium when

---

* In Bailey's experiments, the value used was

$$g = \mu \frac{E}{R}\left[1 - 2\epsilon + (\tfrac{5}{2}m - \epsilon)\cos^2\lambda\right];$$

in which ϵ is the ellipticity of the earth, m the ratio of the centrifugal force at the equator to equatorial gravity, and λ the latitude of the place.

under the action of the masses in their alternate positions, and the time T of oscillation of the torsion rod. To observe these, a small mirror was attached to the rod at C, with its plane nearly perpendicular to the rod. A scale was engraved on a vertical plate at a distance of 108 inches from the mirror, and the image of the scale formed by reflection on the mirror was viewed in a telescope placed just over the scale. In this way an angle of one or two seconds could be read.

The final result was that the mean density of the earth is 5.6747 times that of distilled water at its maximum density.

144. PROBLEM. *If the earth were a homogeneous sphere, at what point in the axis must it be struck, and what momentum must it receive, that it shall have a velocity of translation of V and of rotation of ω?*

Let $M =$ the mass of the sphere,

$R =$ its radius,

$k_1 = \sqrt{\tfrac{2}{5}}R =$ the principal radius of gyration. (See Example 4, page 174), and

$a =$ the distance from the centre to the point where the impulse is applied.

The momentum must be

$$MV,$$

wherever the blow is applied. The moment of an impulse being the same as the moment of the momentum, we have, according to equation (155),

$$\omega = \frac{\text{moment of the momentum}}{\text{moment of inertia}};$$

$$= \frac{MV.a}{Mk_1^2};$$

$$= \frac{Va}{\tfrac{2}{5}R^2};$$

$$\therefore a = \tfrac{2}{5} R^2 \frac{\omega}{V}.$$

The angular velocity of the earth per hour, is

$$\frac{2\pi}{24};$$

and the linear velocity in the orbit is

68,000 miles per hour nearly;

$$\therefore a = \frac{\pi}{2,040,000} R^2.$$

Letting $R = 4,000$ miles, we have

$$a = 24 \text{ miles nearly.}$$

145. PROBLEM. *A homogeneous disc has a motion of translation and of rotation entirely in its own plane, when suddenly any point in the disc becomes fixed; required the angular velocity about the fixed point.*

Let $V =$ the velocity of translation of the centre of the disc;
$\omega =$ the angular velocity about the centre;
$p =$ the perpendicular distance between the fixed point and the line of motion of the centre at the instant that the point becomes fixed;
$r =$ the distance between the fixed point and the centre of the disc;
$k_1 =$ the principal radius of gyration; and
$\omega_1 =$ the angular velocity about the fixed point.

Then

$$\omega_1 = \frac{\text{moment of the momentum}}{\text{moment of inertia}}.$$

$$= \frac{Mk_1^2\omega + MVp}{M(k_1^2 + r^2)};$$

$$= \frac{k_1^2\omega + Vp}{k_1^2 + r^2}.$$

If $V = 0$, we have

$$\omega_1 = \frac{k_1^2\omega}{k_1^2 + r^2}.$$

If the centre becomes fixed, we have $p = 0$, and $r = 0$, and
$$\omega_1 = \omega.$$

146. Problem. *A sphere whose radius is R has an angular velocity ω, and gradually contracts until its radius is r; required the final angular velocity.*

We will assume that the body remains homogeneous throughout, and that there is no change of temperature, and that the change of volume is due simply to the mutual attraction of the particles for each other, which is supposed to draw them towards the centre. Then will the kinetic energy of the body remain constant.

Let ω_1 be the required angular velocity; then we have from equation (153)
$$\tfrac{1}{2} M \cdot \tfrac{2}{5} R^2 \cdot \omega^2 = \tfrac{1}{2} M \cdot \tfrac{2}{5} r^2 \cdot \omega_1^2,$$
$$\therefore \omega_1 = \frac{R}{r} \omega.$$

CHAPTER XI.

GENERAL EQUATIONS OF MOTION.

147. D'Alembert's Principle.—A body is a collection of material particles held together by a force exerted by each particle upon the others. Having deduced the laws of action for forces acting upon a single particle, the direct process for determining the effect of forces upon a body of finite size, appears to be to consider all the forces which act upon each particle separately, including the mutual actions and reactions of the particles, thus establishing equations for each particle of the body, and then to eliminate the terms involving the actions and reactions. But the latter are generally unknown. Various expedients were resorted to by the ancient mathematicians to reach the resulting equations, but the principle announced by D'Alembert greatly simplified the operations, and in many cases reduced the establishment of the equations to Statical principles. See Whewell's History of the Inductive Sciences, Vol. I., p. 365.

The forces which produce the motion of a body may be applied at only a few points, and yet produce motion in every particle of it. These forces are called *impressed forces*. If we consider the particles as separated from each other, and forces applied to them which will produce the same motion that they had when in the body, the latter forces are called *effective forces*. The *effective forces* will then produce the same effect upon the body as the *impressed forces*. D'Alembert's principle consists in this, that *if a system of forces, equal and opposite to the* EFFECTIVE FORCES, *act upon a body, they will be in equilibrium with the* IMPRESSED FORCES.

In this principle no assumption has been made in regard to the character of the mutual actions and reactions between the particles, and hence it is applicable to flexible bodies and fluids, as well as to solids. It is equivalent to assuming that the forces within a body constitute a system which are in equilibrium among themselves.

148. To FIND THE GENERAL EQUATIONS OF MOTION OF A BODY.
Let x_1, y_1, z_1, be the coördinates of a particle whose mass is m_1; X_1, Y_1, Z_1, the impressed forces parallel to the respective axes acting upon the particle; and a similar notation for the other particles. The measure of the accelerating force parallel to the axis of x will be

$$m_1 \frac{d^2 x_1}{dt^2},$$

and if a force equal and opposite to this act upon the particle there will be equilibrium; hence we have

$$X_1 - m_1 \frac{d^2 x_1}{dt^2} = 0; \quad Y_1 - m_1 \frac{d^2 y}{dt^2} = 0; \quad Z_1 - m_1 \frac{d^2 z_1}{dt^2} = 0.$$

Similarly,

$$X_2 - m_2 \frac{d^2 x_2}{dt^2} = 0; \quad Y_2 - m_2 \frac{d^2 y_2}{dt^2} = 0; \quad Z_2 - m_2 \frac{d^2 z_2}{dt^2} = 0;$$

and similar expressions for all the other particles of the body. But if ΣX, ΣY, ΣZ, be the sum of the respective axial components of the *impressed forces*, then

$$\Sigma X = X_1 + X_2 + X_3 + \text{etc.};$$

and similarly for the others. Hence, if m be the mass of any particle whose coördinates are x, y, z, at the time t, we have according to D'Alembert's principle

$$\Sigma X - \Sigma m \frac{d^2 x}{dt^2} = 0;$$

and similarly for the others.

Taking the moments of the forces in reference to the axis of x, we have, in precisely the same way, the equation

$$\Sigma Zy - \Sigma Yz - \Sigma \left(my \frac{d^2 z}{dt^2} - mz \frac{d^2 y}{dt^2} \right) = 0;$$

and similarly for the other axis. Hence, we have the following six equations:—

$$\left.\begin{aligned}\Sigma X - \Sigma m \frac{d^2x}{dt^2} &= 0;\\ \Sigma Y - \Sigma m \frac{d^2y}{dt^2} &= 0;\\ \Sigma Z - \Sigma m \frac{d^2z}{dt^2} &= 0;\end{aligned}\right\} \quad (164)$$

$$\left.\begin{aligned}\Sigma(Zy - Yz) - \Sigma \left(my \frac{d^2z}{dt^2} - mz \frac{d^2y}{dt^2}\right) &= 0;\\ \Sigma(Xz - Zx) - \Sigma \left(mz \frac{d^2x}{dt^2} - mx \frac{d^2z}{dt^2}\right) &= 0;\\ \Sigma(Yx - Xy) - \Sigma \left(mx \frac{d^2y}{dx} - my \frac{d^2x}{dt^2}\right) &= 0.\end{aligned}\right\} \quad (165)$$

Let x_1, y_1, z_1, be the coördinates of any particle of the body referred to a movable system whose origin remains at the centre of the mass, and whose axes are parallel to the fixed axes, and $\bar{x}, \bar{y}, \bar{z}$, the coördinates of the centre of the mass referred to the fixed origin;

then we have

$$x = \bar{x} + x_1;$$
$$\Sigma mx = \Sigma m\bar{x} + \Sigma mx_1;$$
$$\Sigma m \frac{d^2x}{dt^2} = \Sigma m \frac{d^2\bar{x}}{dt^2} + \Sigma m \frac{d^2x_1}{dt^2}.$$

But the origin of the movable system being at the centre of the mass, we have, from equations (71a) or (84a),

$$\Sigma mx_1 = 0;$$
$$\therefore \Sigma m \frac{d^2x_1}{dt^2} = 0;$$

and the last of the preceding equations becomes

$$\Sigma m \frac{d^2x}{dt^2} = \Sigma m \frac{d^2\bar{x}}{dt^2}$$
$$= \frac{d^2\bar{x}}{dt^2} \Sigma m,$$

since $\frac{d^2\bar{x}}{dt^2}$ is a common factor to all the particles m; but $\Sigma m = M$;

$$\therefore \Sigma m \frac{d^2x}{dt^2} = M \frac{d^2\bar{x}}{dt^2},$$

and similarly for the others. Hence, equations (164) become

$$\left. \begin{array}{l} M \dfrac{d^2\bar{x}}{dt^2} = \Sigma X; \\[4pt] M \dfrac{d^2\bar{y}}{dt^2} = \Sigma Y; \\[4pt] M \dfrac{d^2\bar{z}}{dt^2} = \Sigma Z. \end{array} \right\} \qquad (166)$$

Similarly, in equations (165), we have

$$\Sigma my \frac{d^2z}{dt^2} =$$

$$\Sigma \left\{ m \left(\bar{y} + y_1\right) \left(\frac{d^2\bar{z}}{dt^2} + \frac{d^2z_1}{dt^2} \right) \right\} =$$

$$\Sigma m\bar{y}\frac{d^2\bar{z}}{dt^2} + \Sigma m\bar{y}\frac{d^2z_1}{dt^2} + \Sigma my_1\frac{d^2\bar{z}}{dt^2} + \Sigma my_1\frac{d^2z_1}{dt^2}.$$

But $\bar{y}$ and $\bar{z}$ are common factors in their respective terms, therefore the expression becomes

$$\bar{y}\frac{d^2\bar{z}}{dt^2}\Sigma m + \bar{y}\Sigma m\frac{d^2z_1}{dt^2} + \frac{d^2\bar{z}}{dt^2}\Sigma my_1 + \Sigma my_1\frac{d^2z_1}{dt^2};$$

but,

$$\Sigma my_1 = 0; \qquad \Sigma m\frac{d^2z_1}{dt^2} = 0;$$

hence, we finally have

$$\Sigma my\frac{d^2z}{dt^2} = M\bar{y}\frac{d^2\bar{z}}{dt^2} + \Sigma my_1\frac{d^2z_1}{dt^2};$$

and similarly for other terms. In this way the first of equations (165) becomes

$$\Sigma Z y_1 - \Sigma Y z_1 + \Sigma Z\bar{y} - \Sigma Y\bar{z} - \Sigma \left(m y_1 \frac{d^2z_1}{dt^2} - m z_1 \frac{d^2y_1}{dt^2} \right) - M \left(\bar{y}\frac{d^2\bar{z}}{dt^2} - \bar{z}\frac{d^2\bar{y}}{dt^2} \right) = 0.$$

Multiply the third of equations (166) by $\bar{y}$, the second by $\bar{z}$, subtract the latter from the former, and we have

$$M\bar{y}\frac{d^2\bar{z}}{dt^2} - M\bar{z}\frac{d^2\bar{y}}{dt^2} = \Sigma Z\bar{y} - \Sigma Y\bar{z};$$

which, substituted in the preceding equation, gives

$$\Sigma\left(my_1\frac{d^2z_1}{dt^2} - mz_1\frac{d^2y_1}{dt^2}\right) = \Sigma Zy_1 - \Sigma Yz_1.$$

Dropping Σ before X, Y, Z, and letting those letters represent the *total* axial components upon the *entire* body, and $Zy_1 - Yz_1$, etc., the resultant moments of the applied forces, we have the six following equations:

$$\left.\begin{array}{c} M\dfrac{d^2\bar{x}}{dt^2} = X; \\[6pt] M\dfrac{d^2\bar{y}}{dt^2} = Y; \\[6pt] M\dfrac{d^2\bar{z}}{dt^2} = Z; \end{array}\right\} \quad (167)$$

$$\left.\begin{array}{c} \Sigma\left(my_1\dfrac{d^2z_1}{dt^2} - mz_1\dfrac{d^2y_1}{dt^2}\right) = Zy_1 - Yz_1; \\[6pt] \Sigma\left(mz_1\dfrac{d^2x_1}{dt^2} - mx_1\dfrac{d^2z_1}{dt^2}\right) = Xz_1 - Zx_1; \\[6pt] \Sigma\left(mx_1\dfrac{d^2y_1}{dt^2} - my_1\dfrac{d^2x_1}{dt^2}\right) = Yx_1 - Xy_1. \end{array}\right\} \quad (168)$$

Equations (167) do not contain the coördinates of the point of application of the forces, hence, *the motion of translation of the centre of a mass is independent of the point of application of the force or forces;* or, in other words, it is independent of the rotation of the mass.

Equations (168) do not contain the coördinates of the centre of the mass, and being the equations for rotation, show that *the rotation of a mass is independent of the translation of its centre.*

These equations are sufficient for determining all the circumstances of motion of a free solid. In their further use the dashes and subscripts will be omitted.

149. If X, Y, Z, are zero, we have

$$\int \frac{d^2x}{dt^2} = \frac{dx}{dt} + C_1 = 0;$$

and similarly for the others. Transposing, squaring, adding and extracting the square root, give

$$v = \sqrt{\frac{dx^2 + dy^2 + dz^2}{dt^2}} = \sqrt{C_1^2 + C_2^2 + C_3^2}; \quad (169)$$

which, being constant, shows that *the motion of the centre of the mass is rectilinear and uniform.*

This is *the general principle of the* CONSERVATION OF THE CENTRE OF GRAVITY.

CONSERVATION OF AREAS.

150. The expression,

$$ydx - xdy,$$

is, according to Article 112, twice the sectoral area passed over by the radius vector of the body in an instant of time. Hence, if

$$\Sigma(mydx - mxdy) = dA_1;$$

differentiating, we find

$$\Sigma\left(my\frac{d^2x}{dt^2} - mx\frac{d^2y}{dt^2}\right) = \frac{d^2A_1}{dt^2}.$$

If there are no accelerating forces $m\dfrac{d^2x}{dt^2} = 0$; and similarly for the others; hence

$$\frac{d^2A_1}{dt^2} = 0; \quad \frac{d^2A_2}{dt^2} = 0; \quad \frac{d^2A_3}{dt^2} = 0;$$

$$\therefore A_1 = c_1 t; \quad A_2 = c_2 t; \quad A_3 = c_3 t; \quad (170)$$

15

the initial values being zero. These are the projections on the coördinate planes of the areas swept over by the radius vector of the body. They establish the principle of the CONSERVATION OF AREAS. That is,

In any system of bodies, moving without accelerating forces and having only mutual actions upon each other, the projections on any plane of the areas swept over by the radius vector are proportional to the times.

CONSERVATION OF ENERGY.

151. Multiply each of equations (167) by dt, add and integrate, and we have

$$Mv^2 - Mv_0^2 = 2\int(Xdx + Ydy + Zdz);$$

and for a system of bodies, we have

$$\tfrac{1}{2}\Sigma(Mv^2) - \tfrac{1}{2}\Sigma(Mv_0^2) = \Sigma\int(Xdx + Ydy + Zdz). \quad (171)$$

The second member is integrable when the forces are directed towards fixed centres and is a function of the distances between them.

Let a, b, c be the coördinates of one centre,

a_1, b_1, c_1, of another, etc.;

x, y, z, the coördinates of the particle m;

r, r_1, etc., be the distances of the particle from the respective centres;

F, F_1, etc., be the forces directed towards the respective centres;

then, resolving the forces parallel to the axes, we have

$$X = F\cos a + F_1 \cos a + \text{etc.}$$
$$= F\frac{a-x}{r} + F_1\frac{a_1-x}{r} + \text{etc.};$$
$$Y = F\frac{b-y}{r} + F_1\frac{b_1-y}{r} + \text{etc.};$$
$$Z = F\frac{c-z}{r} + F_1\frac{c_1-z}{r} + \text{etc.}$$

Multiplying by dx, dy, dz, respectively, and adding, we have

$$Xdx + Ydy + Zdz = F\left\{\frac{a-x}{r}dx + \frac{b-y}{r}dy + \frac{c-z}{r}dz\right\}$$
$$+ F_1\left\{\frac{a_1-x}{r_1}dx + \frac{b_1-y}{r_1}dy + \frac{c_1-z}{r_1}dz\right\}$$
$$+ \text{etc.}$$

But $\quad r^2 = (a-x)^2 + (b-y)^2 + (c-z)^2;\quad (172)$
and by differentiating, we find

$$dr = -\frac{a-x}{r}dx - \frac{b-y}{r}dy - \frac{c-z}{r}dz;$$

similarly,

$$dr_1 = -\frac{a_1-x}{r_1}dx - \frac{b_1-y}{r_1}dy - \frac{c_1-z}{r_1}dz;$$
$$\text{etc.,}\quad\text{etc.,}\quad\text{etc.}$$

These substituted above, give

$$Xdx + Ydy + Zdz = -Fdr - F_1dr_1 - F_2dr_2 - \text{etc.}$$

Therefore, if F, etc., is a function of r, etc., and μ, μ_1, etc., the intensities of the respective forces at a distance unity from the respective centres, or

$$F = \mu\phi(r);$$
$$F_1 = \mu_1\phi(r_1);$$
$$\text{etc.,}\quad\text{etc.;}$$

the second member, and hence the first, will be integrable.

In nature, if a particle m attracts a particle m_1, the particle m_1 will attract m, each being a centre of force in reference to the other, and *both* centres will be *movable* in reference to a fixed origin. But one centre may be considered fixed in reference to the other, and consequently the proposition remains true for this case.

The second member of equation (171) being integrated between the limits x, y, z and x_0, y_0, z_0, we have

$$\tfrac{1}{2}\Sigma(Mv^2) - \Sigma(Mv_0^2) = \tfrac{1}{2}\Sigma\mu\phi(x,y,z) - \Sigma\mu\phi(x_0,y_0,z_0).\ (173)$$

Hence, *the gain or loss of living force of a system, subject to forces directed towards fixed centres and which are functions of the distances from those centres, is independent of the path* described by the bodies, and depends only upon the position left and arrived at by the bodies, and the intensities of the forces at a unit's distance from the respective centres.

Therefore, when the system returns to the initial position, or to a condition equivalent to the original one, the vis viva will be the same.

From equation (171) we have

The gain or loss of vis viva in passing from one position to another equals twice the work done by the impressed forces.

Let $W_0 =$ the work done by the impressed forces in passing from some definite point (x_0, y_0, z_0) to another definite point (x_1, y_1, z_1); and

$W =$ the work done by the same forces in passing from the first point to any point $(x, y, z,)$ between the two former ones ;

then from equation (171), he have

$$\tfrac{1}{2}\Sigma(Mv_1^2) - \tfrac{1}{2}\Sigma(Mv_0^2) = W_0 ;$$
$$\tfrac{1}{2}\Sigma(Mv_1^2) - \tfrac{1}{2}\Sigma(Mv^2) = W ;$$

and substracting the latter from the former, gives

$$\tfrac{1}{2}\Sigma(Mv^2) - \tfrac{1}{2}\Sigma(Mv_0^2) = W_0 - W ;$$

or, by transposing,

$$\tfrac{1}{2}\Sigma(Mv^2) + W = \tfrac{1}{2}\Sigma(Mv_0^2) + W_0 ;$$

in which the second member is constant.

The first term of the first member is the kinetic energy which the system has at any point of the path, and the second term is the work which has been done by the forces upon the body and has become *latent*, or potential; hence, in such a system *the sum of the kinetic and potential energies is constant*.

This is *the principle of* THE CONSERVATION OF ENERGY in theoretical mechanics. This term has been extended so as to include the principle of the *transmutation of energy* as established by physical science.

If a portion of the universe, as the Solar System for instance, be separated from all external forces, the sum of the kinetic and potential energies will remain constant, so that if the kinetic energy diminishes, the potential increases, and the converse. If external forces act, the potential and kinetic energies may both be increased.

To be more specific, suppose that the earth and sun constitute the system, the sun being considered the centre of the force. The velocity of the earth will be greatest when nearest the sun, and will diminish as it recedes from it. While receding, the amount of work done *against* the attractive force of the sun will be

$$\tfrac{1}{2}Mv^2 - \tfrac{1}{2}Mv_1^2 = -W;$$

in which M is the mass of the earth, v_1 the maximum velocity, and v the velocity at any point. The second member is negative because the first member is.

When the earth is approaching the sun the velocity is increased and the living force is restored, and the kinetic added to the potential energy is constant.

Again, if a body whose weight is W be raised a height h, the work which has been done to raise it to that point is Wh, and in that *position* its potential energy is Wh. If it falls freely through that height it will acquire a velocity $v = \sqrt{2gh}$,

$$\therefore Wh = \frac{W}{2g}v^2 = \tfrac{1}{2}Mv^2$$

which is the kinetic energy.

If the same body fall through a portion of the height, say h_1, its kinetic energy will be $Wh_1 = \tfrac{1}{2}Mv_1^2$, and the work which is still due to its *position* is $W(h-h_1)$, which, at that instant, is *inert* or potential.

It is found, however, that in the use of machines or other devices, by which work is transmitted from one body to another, all the work stored in a moving body cannot be utilized. Thus, in the impact of non-elastic bodies there is always a loss of living force. (See Article 32.)

This, so far as theoretical mechanics is concerned, is a loss, and is treated as such, and until modern Physical Science established *the correlation of forces* it was supposed to be entirely lost. But we know that in the case of impact heat is developed, and Joule determined a definite relation between the quantity of heat and the work necessary to produce it, and called the result *the mechanical equivalent of heat*. Further investigations show that in every case of a supposed loss of energy, it may be accounted for in a general way by the appearance of energy in some other form. It is impossible to trace the transformation of energy as it appears in mechanical action, friction, heat, light, electricity, magnetism, etc., and prove by direct measurements that the sum total at every instant and with every transformation remains rigidly constant; but by means of careful observations and measurements nearly all the energy in a variety of cases has been traced from one mode of action to another, and the small fraction which was apparently lost could be accounted for by the imperfections of the apparatus, or in some other satisfactory manner; until at last the principle of *the conservation of energy* is recognized to be a law as universal as that of the law of gravitation. The *exact nature* of molecular energy which manifests itself in heat, chemical affinity, etc., are unknown, but, according to the general law, all energy whether molecular or of finite masses, is either *kinetic* or *potential;* that is, "Energy of Motion," or "Energy of Position."

COMPOSITION OF ROTARY MOTIONS.

152. *Let the angular velocities about each of the rectangular axes of the movable system be known, it is required to find the instantaneous axis, and the angular velocity.*

Assume any point of the body and observe the path which it describes. Its path at the instant will be perpendicular to the instantaneous axis, and its angular velocity will be the angular velocity required. The path may be oblique to the coördinate planes. Let it be projected on each of the planes, and let *ac* be the projection on *xy*. Let

$$\omega_x, \quad \omega_y, \quad \omega_z,$$

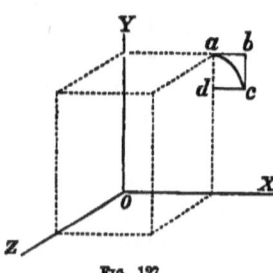

Fig. 127

be the angular velocities about the respective axes. Resolve *ac* parallel to *x* and *y*, *ab* being the former component, and *bc* the latter. Then, at the instant, will *dc* (or its equal *ab*) equal $-\omega_z y$, and $ad = \omega_z x$.

[152.]   ROTARY MOTION.   241

Similarly in regard to the axis of y, we have
$$-\omega_y x, \quad \omega_y z,$$
and in regard to x,
$$-\omega_x z, \quad \omega_x y.$$

When all these rotations take place at the same time, we have, by adding the corresponding velocities, the several velocities along the axes

$$\left. \begin{aligned} \frac{dx}{dt} &= \omega_y z - \omega_z y\,; \\ \frac{dy}{dt} &= \omega_z x - \omega_x z\,; \\ \frac{dz}{dt} &= \omega_x y - \omega_y x. \end{aligned} \right\} \quad (174)$$

The particles on the instantaneous axis have no velocity in reference to the movable origin, hence
$$\frac{dx}{dt} = 0, \quad \frac{dy}{dt} = 0, \quad \frac{dz}{dt} = 0\,;$$
$$\therefore \omega_y z - \omega_z y = 0, \quad \omega_z x - \omega_x z = 0, \quad \omega_x y - \omega_y x = 0\,; \quad (175)$$
which are the equations of a straight line through the origin, and are the equations of the instantaneous axis. Let a, β, γ, be the angles which it makes with the axes x, y, z, respectively, then (*Anal. Geom.*),

$$\cos a = \frac{\omega_x}{\sqrt{\omega_x^2 + \omega_y^2 + \omega_z^2}}\,; \quad \cos \beta = \frac{\omega_y}{\sqrt{\omega_x^2 + \omega_y^2 + \omega_z^2}}\,;$$

$$\cos \gamma = \frac{\omega_z}{\sqrt{\omega_x^2 + \omega_y^2 + \omega_z^2}}.$$

To determine the angular velocity of the body, take any point in a plane perpendicular to the instantaneous axis. Let the point be on the axis of x, and from it erect a perpendicular to the instantaneous axis, and we have

$$p = x \sin a = x\sqrt{1 - \cos^2 a} = \sqrt{\frac{\omega_y^2 + \omega_z^2}{\omega_x^2 + \omega_y^2 + \omega_z^2}}\, x.$$

For this point $y=0$ and $z=0$ in equations (174), and we find for the actual velocity,

$$V = \frac{\sqrt{dx^2 + dy^2 + dz^2}}{dt} = x\sqrt{\omega_y^2 + \omega_z^2};$$

and hence

$$\omega = \frac{V}{p} = \sqrt{\omega_x^2 + \omega_y^2 + \omega_z^2}; \qquad (176)$$

which represents the diagonal of a rectangular parallelopipedon, of which the sides are ω_x, ω_y, ω_z.

153. Moments *of rotation of the centre of the mass about the fixed axes.*

Multiply the second of equations (167) by $\bar{z}$, the third by $\bar{y}$, subtract the former from the latter, and we have

$$M\left(\bar{y}\frac{d^2\bar{z}}{dt^2} - \bar{z}\frac{d^2\bar{y}}{dt^2}\right) = Z\bar{y} - Y\bar{z}.$$

Treating the equations two and two in this manner, dropping the dashes, and substituting L_1, M_1, N_1, for the second members, we have

$$\left.\begin{array}{l} M\left(y\dfrac{d^2z}{dt^2} - z\dfrac{d^2y}{dt^2}\right) = L_1; \\[4pt] M\left(z\dfrac{d^2x}{dt^2} - x\dfrac{d^2z}{dt^2}\right) = M_1; \\[4pt] M\left(x\dfrac{d^2y}{dt^2} - y\dfrac{d^2x}{dt^2}\right) = N_1. \end{array}\right\} \qquad (177)$$

These equations are of the same *form* as equations (168).

MOTION OF A BODY DURING IMPACT.

154. *Motion of the centre of the masses.* The second members of equations (167) are the accelerating forces. If any two of the bodies collide, they being free in other respects, the action of one body upon the other is equal and opposite to that of the latter upon the former; hence, in regard to the system they neutralize each other, and the motion of the centre of the masses will be unaffected by the collision. If there are no accelerating forces the velocity of the centre will be uniform and in a straight line, as shown in Article 140.

To find the velocity of the bodies after impact requires a knowledge of their physical constitution. See Articles 28 and 29.

155. *The motion of rotation* of the *centre* of the entire mass about the origin will also be unaffected by the collision, when the bodies are acted upon by accelerating forces; for, the moments of the forces due to the collision will neutralize each other, and the second members of equations (177) will contain only the applied forces.

This would be illustrated by the impact of two asteroids, or in the bursting of a primary planet.

But the rotation produced in each body about the centre of its mass depends upon the moments of the forces applied to the body, and hence, upon the moment of the momentum produced by the impact.

CONSTRAINED MOTION.

156. *General equations of rotation about a fixed point.*

Take the origin of coördinates at the fixed point. For this case equations (164) vanish. Substitute in (165), the values of $\frac{d^2x}{dt^2}$, etc., obtained from (174), and reduce. We have

$$\frac{d^2x}{dt^2} = z\frac{d\omega_y}{dt} - y\frac{d\omega_z}{dt} + \omega_y(\omega_x y - \omega_y x) - \omega_z(\omega_z x - \omega_x z);$$

and similarly for the others.

Let $L, M, N,$ be substituted for the last term respectively of equations (165), and substituting the above values in the last of these equations, we find

$$\left.\begin{array}{l} \dfrac{d\omega_z}{dt}\Sigma m(x^2+y^2) + \omega_x\omega_y\Sigma m(x^2-y^2) \\[6pt] -(\omega_x^2-\omega_y^2)\Sigma mxy - \left(\dfrac{d\omega_y}{dt}+\omega_x\omega_z\right)\Sigma myz \\[6pt] +\left(\omega_y\omega_z-\dfrac{d\omega_x}{dt}\right)\Sigma mxz. \end{array}\right\} = N. \quad (178)$$

The other two equations may be treated in the same manner. But they are too complicated to be of use. Since the position of the axes is arbitrary, let them be so chosen that

$$\Sigma mxy = 0, \quad \Sigma mxz = 0, \quad \Sigma myz = 0; \quad (179)$$

in which case the axes are called *principal axes ;* and we will show in the next article, that, for every point of a body, there are at least *three* principal axes, each of which is perpendicular to the plane of the other two.

Let x_1, y_1, z_1, be the principal axes, having the same origin as the fixed axes, and

$A = \Sigma m(y_1^2 + z_1^2)$, the moment of inertia of the body about x_1;
$B = \Sigma m(z_1^2 + x_1^2)$, moment about y_1;
$C = \Sigma m(x_1^2 + y_1^2)$, moment about z_1;

also let $\omega_1, \omega_2, \omega_3$, be the angular velocities about the respective axes x_1, y_1, and z_1, and substituting these several quantities in (178), we have

$$\left. \begin{array}{c} A\dfrac{d\omega_1}{dt} + (C - B)\,\omega_2\omega_3 = L; \\[6pt] B\dfrac{d\omega_2}{dt} + (A - C)\,\omega_3\omega_1 = M; \\[6pt] C\dfrac{d\omega_3}{dt} + (B - A)\,\omega_1\omega_2 = N. \end{array} \right\} \quad (180)$$

These are called Euler's Equations.

The origin of coördinates may be taken at the centre of the mass, and as the rotation about that point is the same whether that point be at rest or in motion, as shown at the bottom of page 234, equations (180) are applicable to the rotation of a free body when acted upon by forces in any manner.

PRINCIPAL AXES.

157. *At every point of a body there are at least three principal axes perpendicular to each other.*

When three axes meeting at a point in a body are perpendicular to each other, and so taken that

$$\Sigma mxy = 0, \quad \Sigma myz = 0, \quad \Sigma mzx = 0;$$

they are called Principal Axes.

[157.]   AXES.   245

The planes containing the principal axes are called Principal Planes.

The moments of inertia in reference to the principal axes at any point are called the Principal Moments of Inertia for that point.

Let ON be any line drawn through the origin, making angles a, β, γ, with the respective coördinate axes. First find the moment of inertia about the line ON. From any point of the line ON, erect a perpendicular, NP. The coördinates of P will be x, y, z. Hence we have

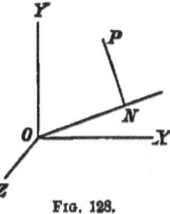

FIG. 128.

$$OP^2 = x^2 + y^2 + z^2;$$
$$ON = x \cos a + y \cos \beta + z \cos \gamma;$$
$$1 = \cos^2 a + \cos^2 \beta + \cos^2 \gamma.$$

The moment of inertia of the body in reference to ON, will be

$$I = \Sigma m \cdot PN^2 = \Sigma m(OP^2 - ON^2)$$
$$= \Sigma m \{ (x^2 + y^2 + z^2) - (x \cos a + y \cos \beta + z \cos \gamma)^2 \}$$
$$= \Sigma m \{ (x^2 + y^2 + z^2)(\cos^2 a + \cos^2 \beta + \cos^2 \gamma) - (x \cos a + y \cos \beta + z \cos \gamma)^2 \}$$
$$= \Sigma m (y^2 + z^2) \cos^2 a + \Sigma m (x^2 + z^2) \cos^2 \beta + \Sigma m (x^2 + y^2) \cos^2 \gamma$$
$$- 2\Sigma myz \cos \beta \cos \gamma - 2\Sigma mzx \cos \gamma \cos a - 2\Sigma mxy \cos a \cos \beta$$
$$= A \cos^2 a + B \cos^2 \beta + C \cos^2 \gamma - 2D \cos \beta \cos \gamma - 2E \cos \gamma \cos a - 2F \cos a \cos \beta;$$

in which A, B, C, have the values given on page 244, and D, E, F, are written for the corresponding factors of the preceding equation.

This may be illustrated geometrically. Conceive a radius vector, r, to move about in space in such a manner that for all angles a, β, γ, corresponding to those of the line ON, the square of the length shall be inversely proportional to the moment of inertia of the body. Then

$$I = \frac{c}{r^2};$$

in which c is a constant. Hence, the polar equation of the locus is

$$\frac{c}{r^2} = A \cos^2 a + B \cos^2 \beta + C \cos^2 \gamma - 2D \cos \beta \cos \gamma - 2E \cos \gamma \cos a - 2F \cos a \cos \beta.$$

Multiplying by r^2, we have

$$c = Ax^2 + By^2 + Cz^2 - 2Dyz - 2Ezx - 2Fxy;$$

which is the equation of the locus referred to rectangular coördinates, and is a quadric. Since $A, B,$ and C are essentially positive, it is the equation of an ellipsoid, and is called *the momental ellipsoid*. Therefore, the moment of inertia about every line which passes through any point of a body may be represented by the radius vector of a certain ellipsoid. But every ellipsoid has at least three principal diameters, hence every material system has, at every point of it, at least three principal axes.

If the ellipsoid be referred to its principal diameters the coefficients of yz, zx, xy, vanish, and the equation of the ellipsoid becomes

$$c = Ax^2 + By^2 + Cz^2.$$

In many cases the principal diameters may be determined by inspection. Thus, in a sphere every diameter is a principal axis. In an ellipsoid the three axes are principal axes. In all surfaces of revolution, the axis of revolution is a principal axis, and any two lines perpendicular to each other and to the axis of revolution are the other two principal axes.

158. *If a body revolve about one of the principal axes passing through the centre of gravity of the body, that axis will suffer no strain from the centrifugal force.*

Let z be a principal axis, about which the body rotates. The centrifugal force of any particle will be

$$m\omega^2 \rho = m\omega^2 \sqrt{x^2+y^2};$$

which, resolved parallel to x and y, gives

$$m\omega^2 x, \quad m\omega^2 y;$$

and the moments of these forces about the axis of z are, for the whole body,

$$\Sigma m\omega^2 xy, \quad \Sigma m\omega^2 yx;$$

but these, according to equations (179), are zero. If the body be free and revolves about this axis it will continue to revolve about it. For this reason it is called *an axis of permanent rotation*.

If the body be free, and the initial rotation be not about a principal axis, the centrifugal force will cause the *instantaneous* axis to change constantly, and it will never rotate about the *permanent axis*. If, therefore, we observe that a free body revolves about an axis for a short time, we infer that it revolved about it from the beginning of the motion.

RELATION BETWEEN THE FIXED AXES x, y, z, AND THE MOVING PRINCIPAL AXES x_1, y_1, z_1.

159. Take the origin of coördinates at the fixed point O, and let it be the centre of a sphere whose radius is unity. Let the line of intersection of the planes xy and x_1y_1 pierce the sphere in the points P, P_1, and let X, X_1, etc., be the points where the respective axes pierce the surface of the sphere. Also, let

$\theta = ZOZ_1$, being the angle between the axes z and z_1, which is also the angle between the planes xy and x_1y_1;

$\phi = POX_1$, being the angle between the line PP_1 and the axis x_1;

$\psi = POX$, being the angle between the line PP_1, and the axis x.

The relation between the parts of the spherical triangles formed by the intersections of the two sets of coördinate planes with the surface of the sphere gives the following results:—

$\cos(z\,z_1) = \cos\theta$;
$\cos(z\,x_1) = -\sin\theta\sin\phi$;
$\cos(z\,y_1) = -\sin\theta\cos\phi$;
$\cos(y\,z_1) = \sin\theta\cos\psi$;
$\cos(y\,y_1) = \sin\phi\sin\psi + \cos\phi\cos\psi\cos\theta$;
$\cos(y\,x_1) = -\cos\phi\sin\psi + \sin\phi\cos\psi\cos\theta$.

160. *Relations between the angular velocities about the several rectangular axes.*

Conceive the body to be rotated a small amount about the axis x_1, then about y_1, and finally about z_1, and that the rotations represent the rates ω_1, ω_2, ω_3. It is required to find the relations between these angular velocities and θ, ϕ, ψ. The arcs described on the sphere by the points X_1, Y_1, and Z_1 will be the angular distances described by the axes.

The new position of the point Z_1 may be considered as having been reached by moving directly away from Z a distance represented by $d\theta$, and then moving perpendicularly away from the plane ZZ_1 on the arc of a small circle, the length of which will be $d\psi \sin \theta$. The rotation of the body about the axis of z_1 will evidently not affect either of these quantities.

The rate of rotation about x_1, projected on the plane zz_1, will be
$$\omega_1 \cos \phi;$$
and, similarly, the rate about y_1, projected on the same plane, will be
$$\omega_2 \sin \phi;$$
and the sum of these will be the rate of increase of θ; hence, we have
$$\frac{d\theta}{dt} = \omega_1 \cos \phi + \omega_2 \sin \phi.$$

Resolving the same rates perpendicularly to the plane zz_1, observing that the resultant rate will be the difference of the components, we have
$$\sin \theta \frac{d\psi}{dt} = -\omega_1 \sin \phi + \omega_2 \cos \phi.$$

The rate of rotation about z_1 will be ω_3, and would equal $\frac{d\theta}{dt}$ if the line PP_1 had not changed its position; but the rotations about the other principal axes has moved that an amount equal to $d\psi$, which, projected on the plane x_1y_1, gives $d\psi \cos \theta$. Hence, we have
$$\frac{d\theta}{dt} = \omega_3 - \frac{d\psi}{dt} \cos \theta.$$

From these we find, by elimination and transformation,

$$\omega_1 = \frac{d\theta}{dt} \cos \phi - \frac{d\psi}{dt} \sin \theta \sin \phi ;$$

$$\omega_2 = \frac{d\theta}{dt} \sin \phi + \frac{d\psi}{dt} \sin \theta \cos \phi ;$$

$$\omega_3 = \frac{d\psi}{dt} \cos \theta + \frac{d\theta}{dt}.$$

The equations in Articles 156, 159, and 160 may be used in the solution of such problems as—The Motion of a Top on a perfectly rough plane, or, in other words, the motion of a top in which the point on which it rests is *fixed* ; —The Motion of a Top on a perfectly smooth horizontal plane;—The Motion of a Gyroscope;—The Motion of the Earth about its centre, considered as an ellipsoid of revolution, and acted upon by the attractive forces of the planets; and other similar problems.

NEW YORK, MARCH, 1877.

TEXT-BOOKS
AND
PRACTICAL WORKS
PUBLISHED BY
JOHN WILEY & SONS,
15 ASTOR PLACE.

Books marked with an * are sold at net prices to the Trade.

AGRICULTURE.

DOWNING. FRUITS AND FRUIT-TREES OF AMERICA; or the Culture, Propagation, and Management in the Garden and Orchard, of Fruit-trees generally, with descriptions of all the finest varieties of Fruit, Native and Foreign, cultivated in this country. By A. J. Downing. Second revision and correction, with large additions. By Chas. Downing. 1 vol. 8vo, over 1100 pages, with several hundred outline engravings. Price, with Supplement for 1872.................................$5 00
"As a work of reference it has no equal in this country, and deserves a place in the Library of every Pomologist in America."—*Marshall P. Wilder.*

" ENCYCLOPEDIA OF FRUITS; or, Fruits and Fruit-Trees of America. Part 1.—APPLES. With an Appendix containing many new varieties, and brought down to 1872. By Chas. Downing. With numerous outline engravings. 8vo, full cloth..$2 50

" ENCYCLOPEDIA OF FRUITS; or, Fruits and Fruit-Trees of America. Part 2.—CHERRIES, GRAPES, PEACHES, PEARS, &c. With an Appendix containing many new varieties, and brought down to 1872. By Chas. Downing. With numerous outline engravings. 8vo, full cloth.........$2 50

* FRUITS AND FRUIT-TREES OF AMERICA. By A. J. Downing. First revised edition. By Chas. Downing. 12mo, cloth...

* SELECTED FRUITS. From Downing's Fruits and Fruit-Trees of America. With some new varieties, including their Culture, Propagation, and Management in the Garden and Orchard, with a Guide to the selection of Fruits, with reference to the Time of Ripening. By Chas. Downing. Illustrated with upwards of four hundred outlines of Apples, Cherries, Grapes, Plums, Pears, &c. 1 vol., 12mo....$2 50

" LOUDON'S GARDENING FOR LADIES, AND COMPANION TO THE FLOWER-GARDEN. Second American from third London edition. Edited by A. J. Downing. 1 vol., 12mo..........................$2 00

DOWNING & THE THEORY OF HORTICULTURE. By J. Lindley.
LINDLEY. With additions by A. J. Downing. 12mo, cloth......$2 00

DOWNING. COTTAGE RESIDENCES. A Series of Designs for Rural Cottages and Cottage Villas, with Garden Grounds. By A. J. Downing. Containing a revised List of Trees, Shrubs, and Plants, and the most recent and best selected Fruit, with some account of the newer style of Gardens. By Henry Winthrop Sargent and Charles Downing. With many new designs in Rural Architecture. By George E Harney, Architect. 1 vol. 4to.................................$6 00

DOWNING & WIGHTWICK.	**HINTS TO PERSONS ABOUT BUILDING IN THE COUNTRY.** By A. J. Downing. And **HINTS TO YOUNG ARCHITECTS**, calculated to facilitate their practical operations. By George Wightwick, Architect. Wood engravings. 8vo, cloth............ ....$2 00
KEMP.	**LANDSCAPE GARDENING**; or, How to Lay Out a Garden. Intended as a general guide in choosing, forming, or improving an estate (from a quarter of an acre to a hundred acres in extent), with reference to both design and execution. With numerous fine wood engravings. By Edward Kemp. 1 vol. 12mo, cloth........................$2 50
LIEBIG	**CHEMISTRY IN ITS APPLICATION TO AGRICULTURE, &c.** By Justus Von Liebig. 12mo, cloth....$1 00
"	**LETTERS ON MODERN AGRICULTURE.** By Baron Von Liebig. Edited by John Blyth, M.D. With addenda by a practical Agriculturist, embracing valuable suggestions, adapted to the wants of American Farmers. 1 vol. 12mo, cloth...$1 00
"	**PRINCIPLES OF AGRICULTURAL CHEMISTRY**, with special reference to the late researches made in England. By Justus Von Liebig. 1 vol. 12mo ....75 cents.
PARSONS.	**HISTORY AND CULTURE OF THE ROSE.** By S. B. Parsons. 1 vol. 12mo $1 25

ARCHITECTURE.

DOWNING.	**COTTAGE RESIDENCES**; or, a Series of Designs for Rural Cottages and Cottage Villas and their Gardens and Grounds, adapted to North America. By A. J. Downing. Containing a revised List of Trees, Shrubs, Plants, and the most recent and best selected Fruits. With some account of the newer style of Gardens, by Henry Wentworth Sargent and Charles Downing. With many new designs in Rural Architecture by George E. Harney, Architect......................$6 00
DOWNING & WIGHTWICK.	**HINTS TO PERSONS ABOUT BUILDING IN THE COUNTRY.** By A. J. Downing. And **HINTS TO YOUNG ARCHITECTS**, calculated to facilitate their practical operations. By George Wightwick, Architect. With many wood-cuts. 8vo, cloth..................$2 00
HATFIELD.	**THE AMERICAN HOUSE CARPENTER.** A Treatise upon Architecture, Cornices, and Mouldings, Framing, Doors, Windows, and Stairs; together with the most important principles of Practical Geometry. New, thoroughly revised, and improved edition, with about 150 additional pages, and numerous additional plates. By R. G. Hatfield. 1 vol. 8vo..$3 50
"	**THE THEORY OF TRANSVERSE STRAINS**, and its Application to the Construction of Buildings, including a full Discussion of the Theory and Construction of Floor Beams, Girders, Headers, Carriage Beams, Bridging, Rolled Iron Beams, Tubular Iron Girders, Cast Iron Girders, Framed Girders and Roof Trusses. With Tables, calculated expressly for this work, of the Dimensions of Floor Beams, Headers, and Rolled Iron Beams; and Tables showing results of original experiments on the Tensile, Transverse and Compressive Strength of American Woods. By R. G. Hatfield, Architect, &c. 8vo, cloth, $3 50
HOLLY	**CARPENTERS' AND JOINERS' HAND-BOOK**, containing a Treatise on Framing Roofs, etc., and useful Rules and Tables. By H. W. Holly. 1 vol. 18mo, cloth...$0 75

JOHN WILEY & SON'S LIST OF PUBLICATIONS.   3

HOLLY	THE ART OF SAW-FILING SCIENTIFICALLY TREATED AND EXPLAINED. With Directions for putting in order all kinds of Saws. By H. W. Holly. 18mo, cloth..$0 75
RUSKIN	SEVEN LAMPS OF ARCHITECTURE. 1 vol. 12mo, cloth, plates..................................$1 75
RUSKIN	LECTURES ON ARCHITECTURE AND PAINTING 1 vol. 12mo, cloth, plates....................$1 50
"	LECTURE BEFORE SOCIETY OF ARCHITECTS. 0 15
WOOD.	A TREATISE ON THE RESISTANCE OF MATERIALS, and an Appendix on the Preservation of Timber By De Volson Wood, Prof. of Engineering, University of Michigan. 2d edition, thoroughly revised. 8vo, cloth.$3 00 This work is used as a Text-Book in Iowa University, Iowa Agricultural College, Illinois Industrial University, Sheffield Scientific School, New Haven, Cooper Institute, New York, Polytechnic College, Brooklyn, University of Michigan, and other institutions.
"	A TREATISE ON BRIDGES. Designed as a Text-book and for Practical Use. By De Volson Wood. 1 vol. 8vo, numerous illustrations, ...$3 00

ASSAYING—ASTRONOMY.

BODEMANN	A TREATISE ON THE ASSAYING OF LEAD, SILVER, COPPER, GOLD, AND MERCURY. By Bodemann and Kerl. Translated by W. A. Goodyear. 1 vol. 12mo, cloth $2 50
MITCHELL.	A MANUAL OF PRACTICAL ASSAYING. By John Mitchell. Fourth edition, edited by William Crookes. 1 vol. thick 8vo, cloth................................$10 00
RICKETTS	NOTES ON ASSAYING AND ASSAY SCHEMES. By P. de Peyster Ricketts, E. M. P. H. D. of School of Mines, Columbia College. 8vo, cloth....................$2 50
NORTON.	A TREATISE ON ASTRONOMY, SPHERICAL AND PHYSICAL, with Astronomical Problems and Solar, Lunar, and other Astronomical Tables for the use of Colleges and Scientific Schools. By William A. Norton. Fourth edition, revised, remodelled, and enlarged. Numerous plates. 8vo, cloth $3 50

BIBLES, &c.

BAGSTER.	THE COMMENTARY WHOLLY BIBLICAL. Contents: —The Commentary: an Exposition of the Old and New Testaments in the very words of Scripture. 2264 pp. II. An outline of the Geography and History of the Nations mentioned in Scripture. III. Tables of Measures, Weights, and Coins. IV. An Itinerary of the Children of Israel from Egypt to the Promised Land. V. A Chronological comparative Table of the Kings and Prophets of Israel and Judah. VI. A Chart of the World's History from Adam to the Third Century, A. D. VII. A complete Series of Illustrative Maps. IX. A Chronological Arrangement of the Old and New Testaments. X. An Index to Doctrines and Subjects, with numerous Selected Passages, quoted in full. XI. An Index to the Names of Persons mentioned in Scripture. XII. An Index to the Names of Places found in Scripture. XIII. The Names, Titles, and Characters of Jesus Christ our Lord, as revealed in the Scriptures, methodically arranged. 3 volumes 4to, cloth.............................. 20 00 3 volumes 4to, half morocco, gilt edges............ 30 00 3 volumes 4to, morocco, gilt edges................ 40 00

JOHN WILEY & SON'S LIST OF PUBLICATIONS.

BLANK-PAGED BIBLE. **THE HOLY SCRIPTURES OF THE OLD AND NEW TESTAMENTS**; with copious references to parallel and illustrative passages, and the alternate pages ruled for MS. notes.

This edition of the Scriptures contains the Authorized Version, illustrated by the references of "Bagster's Polyglot Bible," and enriched with accurate maps, useful tables, and an Index of Subjects.

1 vol. 8vo, morocco extra............................$9 00

1 vol. 8vo, full morocco.............................11 00

BOOK-KEEPING.

JONES. **BOOKKEEPING AND ACCOUNTANTSHIP.** Elementary and Practical. In two parts, with a Key for Teachers. By Thomas Jones, Accountant and Teacher. 1 volume 8vo, cloth..$2 50

" **BOOKKEEPING AND ACCOUNTANTSHIP.** School Edition. By Thomas Jones. 1 vol. 8vo, half roan.......$1 50

" **BOOKKEEPING AND ACCOUNTANTSHIP.** Set of Blanks. In 6 parts. By Thomas Jones..............$1 50

" **BOOKKEEPING AND ACCOUNTANTSHIP.** Double Entry; Results obtained from Single Entry; Equation of Payments, etc. By Thomas Jones. 1 vol. thin 8vo...$0 75

BOTANY.

THOMÉ **STRUCTURAL AND PHYSIOLOGICAL BOTANY.** By Otto W. Thomé, Prof. of Botany at the School of Science and Art, at Cologne. Translated and edited by A. W. Bennett, of St. Thomas Hospital. 600 woodcuts and colored map. Cloth, small 8vo....................$2.25

CHEMISTRY.

CRAFTS. **A SHORT COURSE IN QUALITATIVE ANALYSIS;** with the new notation. By Prof. J. M. Crafts. Second edition. 1 vol. 12mo, cloth......................$1 50

JOHNSON'S FRESENIUS. **A MANUAL OF QUALITATIVE CHEMICAL ANALYSIS.** By C. R. Fresenius. Translated into the New System, and newly edited by Samuel W. Johnson, M.A., Prof. of Theoretical and Agricultural Chemistry, in the Sheffield Scientific School of Yale College, New Haven. 1 vol. 8vo, cloth. 1875...................................$3 50

" **A SYSTEM OF INSTRUCTION IN QUANTITATIVE CHEMICAL ANALYSIS.** By C. R. Fresenius. From latest editions, edited, with additions, by Prof. S. W. Johnson. With Chemical Notation and Nomenclature, old and new...$4 50

KIRKWOOD **COLLECTION OF REPORTS (CONDENSED) AND OPINIONS OF CHEMISTS IN REGARD TO THE USE OF LEAD PIPE FOR SERVICE PIPE,** in the Distribution of Water for the Supply of Cities. By Jas. P. Kirkwood. 8vo, cloth.........................$1 50

MILLER. **ELEMENTS OF CHEMISTRY, THEORETICAL AND PRACTICAL.** By Wm. Allen Miller. 3 vols. 8vo..$18 00

" Part I.—CHEMICAL PHYSICS. 1 vol. 8vo..........$4 00

" Part II.—INORGANIC CHEMISTRY. 1 vol. 8vo..... 6 00

" Part III.—ORGANIC CHEMISTRY. 1 vol. 8vo.......10 00

"Dr. Miller's Chemistry is a work of which the author has every reason to feel proud. It is now by far the largest and most accurately written Treatise on Chemistry in the English language," etc.—*Dublin Med. Journal.*

" **MAGNETISM AND ELECTRICITY.** By Wm. Allen Miller. 1 vol. 8vo...$2 50

JOHN WILEY & SON'S LIST OF PUBLICATIONS. 5

*MUSPRATT. **CHEMISTRY—THEORETICAL, PRACTICAL, AND ANALYTICAL**—as applied and relating to the Arts and Manufactures. By Dr. Sheridan Muspratt. 2 vols. 8vo, cloth, $10.00; half russia............................$25 00

PERKINS. **AN ELEMENTARY MANUAL OF QUALITATIVE CHEMICAL ANALYSIS.** By Maurice Perkins. 12mo, cloth. $1 00

THORPE. **QUANTITATIVE CHEMICAL ANALYSIS.** By T. E. Thorpe, Prof. of Chemistry, Glasgow. 1 vol. 18mo, plates. Cloth..$1 75

Prof. S. W. Johnson says of this work:— "I know of no other small book of anything like its value."

"This very excellent and orginal work has long been waited for by scientific men."—*Scientific American.*

DRAWING, PAINTING AND PERSPECTIVE.

BOUVIER AND OTHERS. **HANDBOOK ON OIL PAINTING.** Handbook of Young Artists and Amateurs in Oil Painting; being chiefly a condensed compilation from the celebrated Manual of Bouvier, with additional matter selected from the labors of Merriwell, De Montalbert, and other distinguished Continental writers on the art. In 7 parts. Adapted for a Text-Book in Academies of both sexes, as well as for self-instruction. Appended, a new Explanatory and Critical Vocabulary. By an American Artist. 12mo, cloth.................$2 00

COE. **PROGRESSIVE DRAWING BOOK.** By Benj. H. Coe. One vol., cloth.......................................$3 50

" **DRAWING FOR LITTLE FOLKS**; or, First Lessons for the Nursery. 30 drawings. Neat cover...........$0 20

" **FIRST STUDIES IN DRAWING.** Containing Elementary Exercises, Drawings from Objects, Animals, and Rustic Figures. Complete in *three numbers* of 18 studies each, in neat covers. Each...............................$0.20

" **COTTAGES.** An Introduction to Landscape Drawing. *Containing 72 Studies.* Complete in four numbers of 18 studies each, in neat covers. Each.....................$0.20

‹ **EASY LESSONS IN LANDSCAPE.** Complete in four numbers of 10 Studies each. In neat 8vo cover. Each, $0 20

" **HEADS, ANIMALS, AND FIGURES.** Adapted to Pencil Drawing. Complete in three numbers of 10 Studies each. In neat 8vo covers. Each................$0 20

" **COPY BOOK, WITH INSTRUCTIONS**............$0 37½

MAHAN. **INDUSTRIAL DRAWING.** Comprising the Description and Uses of Drawing Instruments, the Construction of Plane Figures, the Projections and Sections of Geometrical Solids, Architectural Elements, Mechanism, and Topographical Drawing. With remarks on the method of Teaching the subject. For the use of Academies and Common Schools. By Prof. D. H. Mahan. 1 vol. 8vo. Twenty steel plates. Full cloth..$3 00

RUSKIN. **THE ELEMENTS OF DRAWING.** In Three Letters to Beginners. By John Ruskin. 1 vol. 12mo...........$1 00

" **THE ELEMENTS OF PERSPECTIVE.** Arranged for the use of Schools. By John Ruskin.................$1 00

SMITH. **A MANUAL OF TOPOGRAPHICAL DRAWING.** By Prof. R. S. Smith. New edition with additions. 1 vol. 8vo, cloth, plates...................................$2.00

" **MANUAL OF LINEAR PERSPECTIVE.** Form, Shade, Shadow, and Reflection. By Prof. R. S. Smith. 1 vol. 8vo, plates, cloth......................................$2 00

WARREN. 1. **ELEMENTARY FREE-HAND GEOMETRICAL DRAWING.** A series of progressive exercises on regular lines and forms, including systematic instruction in lettering; a training of the eye and hand for all who are learning to draw. 12mo, cloth, many cuts.....................................75 cts.

DITTO, including *Drafting Instruments*, etc. 12mo, cl..$1 75

JOHN WILEY & SON'S LIST OF PUBLICATIONS.

ELEMENTARY WORKS.—Continued.

WARREN

2. PLANE PROBLEMS IN ELEMENTARY GEOMETRY. With numerous wood-cuts. 12mo, cloth................. $1 25
3. DRAFTING INSTRUMENTS AND OPERATIONS. Containing full information about all the instruments and materials used by the draftsmen, with full directions for their use. With plates and wood-cuts. One vol. 12mo, cloth, $1 25
4. ELEMENTARY PROJECTION DRAWING. Revised and enlarged edition. In five divisions. This and the last volume are favorite text-books, especially valuable to all Mechanical Artisans, and are particularly recommended for the use of all higher public and private schools. New revised and enlarged edition, with numerous wood-cuts and plates. (1872.) 12mo, cloth.. $1 50
5. ELEMENTARY LINEAR PERSPECTIVE OF FORMS AND SHADOWS. Part I.—Primitive Methods, with an Introduction. Part II.—Derivative Methods, with Notes on Aerial Perspective, and many Practical Examples. Numerous wood-cuts. 1 vol. 12mo, cloth........................... $1 00

II. HIGHER WORKS.

These are designed principally for Schools of Engineering and Architecture, and for the members generally of those professions; and the first three are also designed for use in those colleges which provide courses of study adapted to the preliminary general training of candidates for the scientific professions, as well as for those technical schools which undertake that training themselves.

1. DESCRIPTIVE GEOMETRY, OR GENERAL PROBLEMS OF ORTHOGRAPHIC PROJECTIONS. The foundation course for the subsequent theoretical and practical works. 1 vol. 8vo, 24 folding plates and woodcuts............$3 50
2. GENERAL PROBLEMS OF SHADES AND SHADOWS. A wider range of problems than can elsewhere be found in English, and the principles of shading. 1 vol. 8vo, with numerous plates. Cloth.......................... $3 00
3. HIGHER LINEAR PERSPECTIVE. Distinguished by its concise summary of various methods of perspective construction; a full set of standard problems, and a careful discussion of special higher ones. With numerous large plates. 8vo, cloth..$3 50
4. ELEMENTS OF MACHINE CONSTRUCTION AND DRAWING; or, Machine Drawings. With some elements of descriptive and rational cinematics. A Text-Book for Schools of Civil and Mechanical Engineering, and for the use of Mechanical Establishments, Artisans, and Inventors. Containing the principles of gearings, screw propellers, valve motions, and governors, and many standard and novel examples, mostly from present American practice. By S. Edward Warren. 2 vols. 8vo. 1 vol. text and cuts, and 1 vol. large plates... . $7 50

STONE CUTTING. A Treatise on the Graphics and Practice of Stone Cutting, for Engineers, Architects, Masons, and Students. 1 vol. 8vo, plates $2 50

A FEW FROM MANY TESTIMONIALS.

"It seems to me that your Works only need a thorough examination to be introduced and permanently used in all the Scientific and Engineering Schools."
—Prof. J. G. FOX, *Collegiate and Engineering Institute, New York City.*

"I have used several of your Elementary Works, and believe them to be better adapted to the purposes of instruction than any others with which I am acquainted."—H. F. WALLING, *Prof. of Civil and Topographical Engineering, Lafayette College, Easton, Pa.*

"The author has happily divided the subjects into two great portions: the former embracing those processes and problems proper to be taught to all students in Institutions of Elementary Instruction; the latter, those suited to advanced students preparing for technical purposes. The Elementary Books ought to be used in all High Schools and Academies; the Higher ones in Schools of Technology."—WM. W. FOLWELL, *President of University of Minnesota.*

DYEING, &c.

CALVERT. **DYEING AND CALICO PRINTING.** By C. Calvert. Edited by Dr. Stenhouse and C. E. Groves. Illustrated with wood engravings and specimens of printed and dyed fabrics. (Ready in October.) 1 vol. 8vo....................$8 00

MACFARLANE. **A PRACTICAL TREATISE ON DYEING AND CALICO PRINTING.** Including the latest Inventions and Improvements. With an Appendix, comprising definitions of chemical terms, with tables of Weights, Measures, &c. By an experienced Dyer. With a supplement, containing the most recent discoveries in color chemistry. By Robert Macfarlane. 1 vol. 8vo..$5 00

REIMANN **A TREATISE ON THE MANUFACTURE OF ANILINE AND ANILINE COLORS.** By M. Reimann. To which is added the Report on the Coloring Matters derived from Coal Tar, as shown at the French Exhibition, 1867. By Dr. Hofmann. Edited by Wm. Crookes. 1 vol. 8vo, cloth, $2 50

"Dr. Reimann's portion of the Treatise, written in concise language, is profoundly practical, giving the minutest details of the processes for obtaining all the more important colors, with woodcuts of apparatus. Taken in conjunction with Hofmann's Report, we have now a complete history of Coal Tar Dyes, both theoretical and practical."—*Chemist and Druggist*.

ENGINEERING.

AUSTIN **A PRACTICAL TREATISE ON THE PREPARATION, COMBINATION, AND APPLICATION OF CALCAREOUS AND HYDRAULIC LIMES AND CEMENTS.** To which is added many useful recipes for various scientific, mercantile, and domestic purposes. By James G. Austin. 1 vol. 12mo.......................................$2 00

COLBURN **LOCOMOTIVE ENGINEERING AND THE MECHANISM OF RAILWAYS.** A Treatise on the Principles and Construction of the Locomotive Engine, Railway Carriages, and Railway Plant, with examples. Illustrated by Sixty-four large engravings and two hundred and forty woodcuts. By Zerah Colburn. Complete, 20 parts, $15.00; or 2 vols. cloth..$16 00
Or, half morocco, gilt top........................$20 00

DU BOIS. **ELEMENTS OF GRAPHICAL STATICS**, and their Application to Framed Structures, etc. Cranes; Bridge, Roof, and Suspension Trusses; Braced and Stone Arches; Pivot and Draw Spans; Continuous Girders, etc. By A. J. Du Bois, C.E., Ph.D. 2 vols. 8vo, 1 vol. text and 1 vol. plates...$5 00

" **HYDRAULICS AND HYDRAULIC MOTORS.** Translated from Vol. II. Weisbach's Mechanics. By Prof. A. Jay Du Bois. 1 vol. 8vo, illustrated.

" **THEORY OF STEAM ENGINE.** Translated from Vol. II. Weisbach's Mechanics. By A. J. Du Bois. 1 vol. 8vo, illustrated.

" **IRON AND STEEL, CALCULATIONS OF STRENGTH AND DIMENSIONS OF.** Translated from Prof. Jacob Weyrauth's Work. By Prof. A. Jay Du Bois. 1 vol. 8vo, illustrated.

HERSCHEL **A HANDBOOK FOR BRIDGE ENGINEERS.** By C. Herschel. In 3 vols. Each vol. complete in itself. Vol. I. Straight and Beam Bridges. Vol. II. Suspension and Arched Bridges. Vol. III. Stone Bridges; Bridge Piers and their Foundations.

MAHAN. **AN ELEMENTARY COURSE OF CIVIL ENGINEERING**, for the use of the Cadets of the U. S. Military Academy. By D. H. Mahan. 1 vol. 8vo, with numerous illustrations, and an Appendix and general Index. Edited by Prof. De Volson Wood. Full cloth........................$5 00

JOHN WILEY & SON'S LIST OF PUBLICATIONS.

MAHAN   **DESCRIPTIVE GEOMETRY**, as applied to the Drawing of Fortifications and Stone Cutting. For the use of the Cadets of the U. S. Military Academy. By Prof. D. H. Mahan. 1 vol. 8vo. Plates........................$1 50

"   **A TREATISE ON FIELD FORTIFICATIONS.** Containing instructions on the Methods of Laying out, Constructing, Defending, and Attacking Entrenchments. With the General Outlines, also, of the Arrangement, the Attack, and Defence of Permanent Fortifications. By Prof. D. H. Mahan. New edition, revised and enlarged. 1 vol. 8vo, full cloth, with plates..............................$3 50

"   **ELEMENTS OF PERMANENT FORTIFICATIONS.** By Prof. D. H. Mahan. 1 vol. 8vo, with numerous large plates. Revised and edited by Col. J. B. Wheeler...$6 50

MAHAN.   **ADVANCED GUARD, OUT-POST**, and Detachment Service of Troops, with the Essential Principles of Strategy and Grand Tactics. For the use of Officers of the Militia and Volunteers. By Prof. D. H. Mahan. New edition, with large additions and 12 plates. 1 vol. 18mo, cloth......$1 50

MAHAN & MOSELY.   **MECHANICAL PRINCIPLES OF ENGINEERING AND ARCHITECTURE.** By Henry Mosely, M.A., F.R.S. From last London edition, with considerable additions, by Prof. D. H. Mahan, LL.D., of the U. S. Military Academy. 1 vol. 8vo, 700 pages. With numerous cuts. Cloth...$5 00

MAHAN & BRESSE.   **HYDRAULIC MOTORS.** Translated from the French Cours de Mecanique, appliquée par M. Bresse. By Lieut. F. A. Mahan, and revised by Prof. D. H. Mahan. 1 vol. 8vo, plates. New Edition, 1876........................$2 50

WOOD.   **A TREATISE ON THE RESISTANCE OF MATERIALS,** and an Appendix on the Preservation of Timber. By De Volson Wood, Professor of Engineering, University of Michigan. 1 vol. 8vo, cloth..........................$3 00

A TREATISE ON BRIDGES. Designed as a Text-book and for Practical Use. By De Volson Wood. 1 vol. 8vo, numerous illustrations, cloth $3 00

' GEOMETRY.

MAHAN   **DESCRIPTIVE GEOMETRY.** As applied to the Drawing of Fortifications and Stone Cutting. By Prof. D. H. Mahan. 8vo, plates, cloth......................$1 50

SEARLE   **ELEMENTS OF GEOMETRY.** By G. M. Searle, C. S. P., formerly Assistant Professor U. S. Naval Academy, &c. 8vo, cloth...

WARREN   **DESCRIPTIVE GEOMETRY, OR GENERAL PROBLEMS OF ORTHOGRAPHIC PROJECTIONS.** 1 vol. 8vo, 24 folding plates, cloth....................$3 50

"   **PLANE PROBLEMS ON ELEMENTARY GEOMETRY.** 12mo.............................. 1 25

GREEK.

BAGSTER   **GREEK TESTAMENTS, ETC.** The Critical Greek and English New Testament in Parallel Columns, consisting of the Greek Text of Scholz, readings of Griesbach, etc., etc. 1 vol. 18mo, half morocco...................:$2 50

"   ——do.   Full morocco, gilt edges.......... 4 50

"   ——With Lexicon, by T. S. Green. Half bound...... 4 00

"   —— do. Full morocco, gilt edges................ 6 00

GREEK AND ENGLISH TESTAMENT. Lexicon and Concordance. Half bound.........................$5 00

"   —— Morocco limp, $6.50; morocco flaps, $7.00; morocco, projecting edges, calf lined................ 7 50

JOHN WILEY & SON'S LIST OF PUBLICATIONS.   9

BAGSTER	**THE ANALYTICAL GREEK LEXICON TO THE NEW TESTAMENT.** In which, by an alphabetical arrangement, is found every word in the Greek text *in every form in which it appears*—that is to say, every occurrent person, number, tense or mood of verbs, every case and number of nouns, pronouns, &c., and is placed in its alphabetical order, fully explained by a careful grammatical analysis and referred to its root. 1 vol. small 4to, half bound,.........................$6 50
"	**GREEK TESTAMENT.** By Griesbach and Greenfield. 32mo. Half bound...............................$1 75
"	DITTO. With Lexicon. 32mo, half bound...........$2 25
GREENFIELD	**GREEK LEXICON.** (Polymicrian). 32mo, half bound, $1 00
GREEN	**GREEK-ENGLISH LEXICON TO TESTAMENT.** By T. S. Green. Half morocco........................$1 50

HEBREW AND CHALDEE.

GREEN	**A GRAMMAR OF THE HEBREW LANGUAGE.** With copious Appendixes. By W. H. Green, D.D., Prof. in Princeton Theological Seminary. 1 vol. 8vo, cloth, $3 50
"	**AN ELEMENTARY HEBREW GRAMMAR.** With Tables, Reading Exercises, and Vocabulary. By Prof. W. H. Green, D.D. 1 vol. 12mo, Cloth................$1 25
"	**HEBREW CHRESTOMATHY**; or, Lessons in Reading and Writing Hebrew. By Prof. W. H. Green, D.D. 1 vol. 8vo, cloth..$2 00
*LETTERIS	**A NEW AND BEAUTIFUL EDITION OF THE HEBREW BIBLE.** Revised and carefully examined by Myer Levi Letteris. 1 vol. 8vo, with key, marble edges. $2 50
LUZZATTO	**GRAMMAR OF THE BIBLICAL CHALDAIC LANGUAGE AND THE TALMUD BABLI IDIOMS.** By S. D. Luzzatto. Translated by Dr. J. S. Goldammer, Rabbi. 1 vol. 12mo, cloth........................$1 50
BAGSTER'S GESENIUS	**BAGSTER'S COMPLETE EDITION OF GESENIUS' HEBREW AND CHALDEE LEXICON.** In large, clear and perfect type. Translated and edited with additions and corrections, by S. P. Tregelles, LL.D. Small 4to, half bound............................$7 00
*BAGSTER'S.	**ANALYTICAL HEBREW AND CHALDEE LEXICON** With an Alphabetical Arrangement of every Word in Old Testament, &c., &c. By B. Davidson. 1 vol. small 4to, half-bound....................................$11.00
	NEW POCKET HEBREW AND ENGLISH LEXICON The arrangement of this Manual Lexicon combines two things—the etymological order of roots and the alphabetical order of words. This arrangement tends to lead the learner onward; for, as he becomes more at home with roots and derivatives, he learns to turn at once to the root, without first searching for the particular word in its alphabetic order. 1 vol. 18mo, cloth....................................$2 00 "This is the most beautiful, and at the same time the most correct and perfect Manual Hebrew Lexicon we have ever used."—*Eclectic Review*.

IRON, METALLURGY, &c.

BODEMANN.	**A TREATISE ON THE ASSAYING OF LEAD, SILVER, COPPER, GOLD, AND MERCURY.** By Bodemann & Kerl. Translated by W. A. Goodyear. 1 vol. 12mo, $2 50
CROOKES.	**A PRACTICAL TREATISE ON METALLURGY.** Adapted from the last German edition of Prof. Kerl's Metallurgy. By William Crookes and Ernst Rohrig. In three vols. thick 8vo. Price..$30 00 Separately. Vol. 1. Lead, Silver, Zinc, Cadmium, Tin, Mercury, Bismuth, Antimony, Nickel, Arsenic, Gold, Platinum, and Sulphur....................................$10 00 Vol. 2. Copper and Iron............................ 10 00 Vol. 3. Steel, Fuel, and Supplement................ 10 00

DUNLAP.	**WILEY'S AMERICAN IRON TRADE MANUAL** of the leading Iron Industries of the United States. With a description of the Blast Furnaces, Rolling Mills, Bessemer Steel Works, Crucible Steel Works, Car Wheel and Car Works, Locomotive Works, Steam Engine and Machine Works, Iron Bridge Works, Stove Foundries, &c., giving their location and capacity of product. With some account of Iron Ores. By Thomas Dunlap, of Philadelphia. 1 vol. 4to. Price to subscribers $7 50
FAIRBAIRN	**CAST AND WROUGHT IRON FOR BUILDING.** By Wm. Fairbairn. 8vo, cloth.........................$2 00
FRENCH.	**HISTORY OF IRON TRADE, FROM 1621 TO 1857.** By B. F. French. 8vo, cloth.........................$2 00
KIRKWOOD	**COLLECTION OF REPORTS (CONDENSED) AND OPINIONS OF CHEMISTS IN REGARD TO THE USE OF LEAD PIPE FOR SERVICE PIPE**, in the Distribution of Water for the Supply of Cities. By I. P. Kirkwood, C.E. 8vo, cloth.......................$1 50
SVEDELIUS	**HAND-BOOK FOR CHARCOAL BURNERS.** Translated from the Swedish by Prof. R. B. Anderson, and edited by Prof. W. J. L. Nicodemus, C. E. 1 vol., 12mo. Plates. Cloth, ...$1.50
WEYRAUTH	**IRON AND STEEL, STRENGTH AND DIMENSIONS, CALCULATIONS OF.** By Prof. Jacob Weyrauth. Translated by Prof. A. Jay Du Bois. 8vo. Plates.

MACHINISTS, &C.

FITZGERALD	**THE BOSTON MACHINIST.** A complete School for the Apprentice and Advanced Machinist. By W. Fitzgerald. 1 vol. 18mo, cloth...................................$0 75
HOLLY.	**SAW FILING.** The Art of Saw Filing Scientifically Treated and Explained. With Directions for putting in order all kinds of Saws, from a Jeweller's Saw to a Steam Saw-mill. Illustrated by forty-four engravings. Third edition. By H. W. Holly. 1 vol. 18mo, cloth........................$0 75
TURNING, &c.	**LATHE, THE, AND ITS USES, ETC.;** or, Instruction in the Art of Turning Wood and Metal. Including a description of the most modern appliances for the ornamentation of plane and curved surfaces, with a description also of an entirely novel form of *Lathe* for Eccentric and Rose Engine Turning, a Lathe and Turning Machine combined, and other valuable matter relating to the Art. 1 vol. 8vo, copiously illustrated. Including Supplement. 8vo, cloth......$7 00 "The most complete work on the subject ever published."—*American Artisan.* "Here is an invaluable book to the practical workman and amateur."—*London Weekly Times.*

MANUFACTURES.

BOOTH.	**NEW AND COMPLETE CLOCK AND WATCH MAKERS' MANUAL.** Comprising descriptions of the various gearings, escapements, and Compensations now in use in French, Swiss, and English clocks and watches, Patents, Tools, etc., with directions for cleaning and repairing. With numerous engravings. Compiled from the French, with an Appendix containing a History of Clock and Watch Making in America. By Mary L. Booth. With numerous plates. 1 vol. 12mo, cloth...................................$2 00
GELDARD.	**HANDBOOK ON COTTON MANUFACTURE;** or, A Guide to Machine-Building, Spinning, and Weaving. With practical examples, all needful calculations, and many useful and important tables. The whole intended to be a complete yet compact authority for the manufacture of cotton. By James Geldard. With steel engravings. 1 vol. 12mo, cloth..$2 50

MECHANICS.

MAGNUS — **LESSONS IN ELEMENTARY MECHANICS.** Introductory to the study of Physical Science. Designed for the use of Schools, etc. By Philip Magnus. With Emendations and Preface by Prof. De Volson Wood. With numerous examples and 121 wood engravings, 18mo, cloth..$1.50

WILLIS — **PRINCIPLES OF MECHANISM.** Designed for the use of Students in the Universities and for Engineering Students generally. By Robert Willis, M.D., F.R.S., President of the British Association for the Advancement of Science, &c., &c. Second edition, enlarged. 1 vol. 8vo, cloth..$7 50

*** It ought to be in every large Machine Workshop Office, in every School of Mechanical Engineering at least, and in the hands of every Professor of Mechanics, &c.—Prof. S. EDWARD WARREN.

WOOD — **THE ELEMENTS OF ANALYTICAL MECHANICS.** With numerous examples and illustrations. For use in Scientific Schools and Colleges. By Prof. De Volson Wood. With numerous wood engravings. 8vo, cloth, $3 00

" — **PRINCIPLES OF ELEMENTARY MECHANICS.** Fully illustrated. 12mo, cloth. (*In preparation.*)

MEDICAL, &c.

BULL. — **HINTS TO MOTHERS FOR THE MANAGEMENT OF HEALTH DURING THE PERIOD OF PREGNANCY, AND IN THE LYING-IN ROOM.** With an exposure of popular errors in connection with those subjects. By Thomas Bull, M.D. 1 vol. 12mo, cloth..........$1 00

FRANCKE — **OUTLINES OF A NEW THEORY OF DISEASE,** applied to Hydropathy, showing that water is the only true remedy. With observations on the errors committed in the practice of Hydropathy, notes on the cure of cholera by cold water, and a critique on Priessnitz's mode of treatment. Intended for popular use. By the late H. Francke. Translated from the German by Robert Blakie, M.D. 1 vol. 12mo, cloth...$1 50

GREEN — **A TREATISE ON DISEASES OF THE AIR PASSAGES.** Comprising an inquiry into the History, Pathology, Causes, and Treatment of those Affections of the Throat called Bronchitis, Chronic Laryngitis, Clergyman's Sore Throat, etc., etc. By Horace Green, M.D. Fourth edition, revised and enlarged. 1 vol. 8vo, cloth...................................$3 00

" — **A PRACTICAL TREATISE ON PULMONARY TUBERCULOSIS,** embracing its History, Pathology, and Treatment. By Horace Green, M.D. Colored plates. 1 vol. 8vo, cloth..$5 00

GREEN — **OBSERVATIONS ON THE PATHOLOGY OF CROUP** With Remarks on its Treatment by Topical Medications. By Horace Green, M.D. 1 vol. 8vo, cloth..............$1 25

" — **ON THE SURGICAL TREATMENT OF POLYPI OF THE LARYNX, AND ŒDEMA OF THE GLOTTIS.** By Horace Green, M.D. 1 vol. 8vo.................$1 25

" — **FAVORITE PRESCRIPTIONS OF LIVING PRACTITIONERS.** With a Toxicological Table, exhibiting the Symptoms of Poisoning, the Antidotes for each Poison, and the Test proper for their detection. By Horace Green. 1 vol. 8vo, cloth................................$2 50

VON DUBEN. — **GUSTAF VON DUBEN'S TREATISE ON MICROSCOPICAL DIAGNOSIS.** With 71 engravings. Translated, with additions, by Prof. Louis Bauer, M.D. 1 vol. 8vo, cloth.....................................$1 00

MINERALOGY.

BRUSH — **MANUAL OF DETERMINATIVE MINERALOGY**, with an Introduction on Blow-Pipe Analysis, being the Determinative Portion of Dana's Mineralogy. By Prof. Geo. J. Brush. 1 vol. 8vo.................................$3 00

DANA — **A SYSTEM OF MINERALOGY.** Descriptive Mineralogy. Comprising the most recent Discoveries. Fifth Edition. Almost entirely re-written and greatly enlarged. Containing nearly 900 pages 8vo, and upwards of 600 wood engravings. By Prof. J. Dana. Cloth.............$10 00

DANA & BRUSH — **APPENDIXES TO DANA'S MINERALOGY**, bringing the work down to 1875. 8vo.......................$1 00

DANA — **A TEXT-BOOK OF MINERALOGY.** After the plan of and with the cooperation of Prof. Jas. D. Dana, of Yale College. Embracing a full Treatise upon Crystallography and Physical Mineralogy, by Edward S. Dana, Ph.D., Curator of Mineralogy, Yale College. With upwards of 800 wood cuts, and a colored plate. 8vo, cloth......$5 00

SHIP-BUILDING, &c.

BOURNE. — **A TREATISE ON THE SCREW PROPELLER, SCREW VESSELS, AND SCREW ENGINES**, as adapted for Purposes of Peace and War. Illustrated by numerous woodcuts and engravings. By John Bourne. New edition. 1867. 1 vol. 4to, cloth, $18.00; half russia................$24 00

WATTS. — **RANKINE (W. J. M.) AND OTHERS.** Ship-Building, Theoretical and Practical, consisting of the Hydraulics of Ship-Building, or Buoyancy, Stability, Speed and Design—The Geometry of Ship-Building, or Modelling, Drawing, and Laying Off—Strength of Materials as applied to Ship-Building —Practical Ship-Building—Masts, Sails, and Rigging—Marine Steam Engineering—Ship-Building for Purposes of War. By Isaac Watts, C.B., W. J. M. Rankine, C.B., Frederick K. Barnes, James Robert Napier, etc. Illustrated with numerous fine engravings and woodcuts. Complete in 30 numbers, boards, $35.00; 1 vol. folio, cloth, $37.50; half russia, $40 00

WILSON (T. D.) **SHIP-BUILDING, THEORETICAL AND PRACTICAL.** In Five Divisions.—Division I. Naval Architecture. II. Laying Down and Taking off Ships. III. Ship-Building IV. Masts and Spar Making. V. Vocabulary of Terms used—intended as a Text-Book and for Practical Use in Public and Private Ship-Yards. By Theo. D. Wilson, Assistant Naval Constructor, U. S. Navy; Instructor of Naval Construction, U. S. Naval Academy; Member of the Institution of Naval Architects, England. With numerous plates, lithographic and wood. 1 vol. 8vo. $7 50

SOAP.

MORFIT. — **A PRACTICAL TREATISE ON THE MANUFACTURE OF SOAPS.** With numerous wood-cuts and elaborate working drawings. By Campbell Morfit, M.D., F.C.S. 1 vol. 8vo...$20 00

STEAM ENGINE.

DU BOIS — **THEORY OF THE STEAM ENGINE.** Translated from Vol. II. Weisbach's Mechanics. By Prof. A. J. Du Bois.

TROWBRIDGE — **TABLES, WITH EXPLANATIONS, OF THE NON-CONDENSING STATIONARY STEAM ENGINE**, and of High-Pressure Steam Boilers. By Prof. W. P. Trowbridge, of Yale College Scientific School. 1 vol. 4to, plates..$2 50

" — **HEAT AS A SOURCE OF POWER**: with applications of general principles to the construction of Steam Generators. An introduction to the study of Heat Engines. By W. P. Trowbridge, Prof. Sheffield Scientific School, Yale College. Profusely illustrated. 1 vol. 8vo, cloth, $3 50

TEXT-BOOKS for Use of U. S. Naval Academy.

COOKE. **A TEXT-BOOK OF NAVAL ORDNANCE AND GUNNERY.** Prepared for the use of the Cadet Midshipmen at the United States Naval Academy. By A. P. Cooke, Com. U. S. N. One thick volume, illustrated by about 400 fine cuts. Cloth................................$12.50

RICE & JOHNSON. **ELEMENTS OF THE DIFFERENTIAL CALCULUS,** founded on the Method of Rates or Fluxions. 8vo.

WILSON. **SHIP-BUILDING, THEORETICAL AND PRACTICAL.** By T. D. Wilson. (See page 15.) 8vo, cloth........$7.50

TURNING, &c.

THE LATHE, **AND ITS USES, ETC.** On Instructions in the Art of Turning Wood and Metal. Including a description of the most modern appliances for the ornamentation of plane and curved surfaces. With a description, also, of an entirely novel form of Lathe for Eccentric and Rose Engine Turning, a Lathe and Turning Machine combined, and other valuable matter relating to the Art. 1 vol. 8vo, copiously illustrated, cloth..........$7 00

" **SUPPLEMENT AND INDEX TO SAME.** Paper...$0 90

VENTILATION.

LEEDS (L. W.). **A TREATISE ON VENTILATION.** Comprising Seven Lectures delivered before the Franklin Institute, showing the great want of improved methods of Ventilation in our buildings, giving the chemical and physiological process of respiration, comparing the effects of the various methods of heating and lighting upon the ventilation, &c. Illustrated by many plans of all classes of public and private buildings, showing their present defects, and the best means of improving them. By Lewis W. Leeds. 1 vol. 8vo, with numerous wood-cuts and colored plates. Cloth........$2 50

"It ought to be in the hands of every family in the country."—*Technologist.*
"Nothing could be clearer than the author's exposition of the principles of the principles and practice of both good and bad ventilation."—*Van Nostrand's Engineering Magazine.*
"The work is every way worthy of the widest circulation."—*Scientific American.*

REID. **VENTILATION IN AMERICAN DWELLINGS.** With a series of diagrams presenting examples in different classes of habitations. By David Boswell Reid, M.D. To which is added an introductory outline of the progress of improvement in ventilation. By Elisha Harris, M.D. 1 vol. 12mo, $1 50

J. W. & SONS are Agents for and keep in stock

SAMUEL BAGSTER & SONS' PUBLICATIONS,

LONDON TRACT SOCIETY PUBLICATIONS,

MURRAY'S TRAVELLER'S GUIDES,

WEALE'S SCIENTIFIC SERIES.

Full Catalogues gratis on application.

J. W. & SONS import to order, for the TRADE AND PUBLIC.

BOOKS, PERIODICALS, &c.,
FROM
ENGLAND, FRANCE, AND GERMANY.

⁎⁎⁎ JOHN WILEY & SONS' Complete Classified Catalogues of the most valuable and latest scientific publications, Parts I. and II., 8vo, mailed to order on the receipt of 10 cts.

RUSKIN'S WORKS.
Uniform in size and style.

RUSKIN — MODERN PAINTERS. 5 vols. tinted paper, bevelled boards, plates, in box.......................................$18 00

" — MODERN PAINTERS. 5 vols. half calf............ 27 00

"   "   "   " without plates....... 12 00

"   "   "   "   half calf, 20 00

Vol. 1.—Part 1. General Principles. Part 2. Truth.
Vol. 2.—Part 3. Of Ideas of Beauty.
Vol. 3.—Part 4. Of Many Things.
Vol. 4.—Part 5. Of Mountain Beauty.
Vol. 5.—Part 6. Leaf Beauty. Part 7. Of Cloud Beauty. Part 8. Ideas of Relation of Invention, Formal. Part 9. Ideas of Relation of Invention, Spiritual.

" — STONES OF VENICE. 3 vols., on tinted paper, bevelled boards, in box...............................$7 00

" — STONES OF VENICE. 3 vols., on tinted paper, half calf..$12 00

" — STONES OF VENICE. 3 vols., cloth............... 7 00
Vol. 1.—The Foundations.
Vol. 2.—The Sea Stories.
Vol. 3.—The Fall.

" — SEVEN LAMPS OF ARCHITECTURE. With illustrations, drawn and etched by the author. 1 vol. 12mo, cloth, $1 75

" — LECTURES ON ARCHITECTURE AND PAINTING. With illustrations drawn by the author. 1 vol. 12mo, cloth....................................$1 50

" — THE TWO PATHS. Being Lectures on Art, and its Application to Decoration and Manufacture. With plates and cuts. 1 vol. 12mo, cloth.......................$1 25

" — THE ELEMENTS OF DRAWING. In Three Letters to Beginners. With illustrations drawn by the author. 1 vol. 12mo, cloth..................................$1 00

" — THE ELEMENTS OF PERSPECTIVE. Arranged for the use of Schools. 1 vol. 12mo, cloth..................$1 00

" — THE POLITICAL ECONOMY OF ART. 1 vol. 12mo, cloth...$1 00

" — PRE-RAPHAELITISM.
NOTES ON THE CONSTRUCTION OF SHEEPFOLDS.
KING OF THE GOLDEN RIVER; or, The Black Brothers. A Legend of Stiria.
} 1 vol. 12mo, cloth, $1 00

RUSKIN — SESAME AND LILIES. Three Lectures on Books, Women, &c. 1. Of Kings' Treasuries. 2. Of Queens' Gardens. 3. Of the Mystery of Life. 1 vol. 12mo, cloth.........$1 50

" — AN INQUIRY INTO SOME OF THE CONDITIONS AT PRESENT AFFECTING "THE STUDY OF ARCHITECTURE" IN OUR SCHOOLS. 1 vol. 12mo, paper....................................$0 15

" — THE ETHICS OF THE DUST. Ten Lectures to Little Housewives, on the Elements of Crystallization. 1 vol. 12mo, cloth..................................$1 25

" — "UNTO THIS LAST." Four Essays on the First Principles of Political Economy. 1 vol. 12mo, cloth.. ....$1 00

JOHN WILEY & SON'S LIST OF PUBLICATIONS. 15

RUSKIN THE CROWN OF WILD OLIVE. Three Lectures on Work, Traffic, and War. 1 vol. 12mo. cloth..............$1 00
" TIME AND TIDE BY WEARE AND TYNE. Twenty-five Letters to a Workingman on the Laws of Work. 1 vol. 12mo, cloth.....................................$1 00
" THE QUEEN OF THE AIR. Being a Study of the Greek Myths of Cloud and Storm. 1 vol. 12mo, cloth $1 0C
" LECTURES ON ART. 1 vol. 12mo, cloth............ 1 00
" FORS CLAVIGERA. Letters to the Workmen and Labourers of Great Britain. Part 1. 1 vol. 12mo, cloth, plates, $1 00
" FORS CLAVIGERA. Letters to the Workmen and Labourers of Great Britain. Part 2. 1 vol. 12mo, cloth, plates, $1 00
" MUNERA PULVERIS. Six Essays on the Elements of Political Economy. 1 vol. 12mo, cloth..............$1 00
" ARATRA PENTELICI. Six Lectures on the Elements of Sculpture, given before the University of Oxford. By John Ruskin. 12mo, cloth, $1 50, or with plates.........$3 00
" THE EAGLE'S NEST. Ten Lectures on the relation of Natural Science to Art. 1 vol. 12mo..............$1 50
" THE POETRY OF ARCHITECTURE: Villa and Cottage. With numerous plates. By Kata Phusin. 1 vol. 12mo, cloth...$1 50
Kata Phusin is the supposed Nom de Plume of John Ruskin.
" FORS CLAVIGERA. Letters to the Workmen and Laborers of great Britain. Part 3. 1 vol. 12mo, cloth........
" LOVES MEINE. Lectures on Greek and English Birds. By John Ruskin. Plates, cloth......................$0 75
" ARIADNE FLORENTINA. Lectures on Wood and Metal Engraving. By John Ruskin, Cloth............$1 50
" FRONDE'S AGRESTES. Readings on "Modern Painters." Chosen at her pleasure by the author's friend, the Younger Lady of the Thwaite, Coniston. 1 vol. 12mo, cloth, $1 00
" THE TRUE AND THE BEAUTIFUL IN NATURE, ART, MORALS, AND RELIGION. Selected from the Works of John Ruskin, A.M. With a notice of the author by Mrs. L. C. Tuthill. Portrait. 1 vol. 12mo, cloth, plain, $2.00; cloth extra, gilt head........ ....$2 5(
" ART CULTURE. Consisting of the Laws of Art selected from the Works of John Ruskin, and compiled by Rev. W. H Platt. A beautiful volume, with many illustrations. 1 vol. 12mo, cloth, extra gilt head......................$3 00
" Do. Do. School edition. 1 vol. 12mo, plates, cloth..$2 50
" PRECIOUS THOUGHTS: Moral and Religious. Gathered from the Works of John Ruskin, A.M. By Mrs. L. C. Tuthill. 1 vol. 12mo, cloth, plain, $1.50. Extra cloth, gilt head..$2 00
" SELECTIONS FROM THE WRITINGS OF JOHN RUSKIN. 1 vol. 12mo, cloth, plain, $2.00. Extra cloth, gilt head..$2 5(
" SESAME AND LILIES. 1 vol. 12mo, ex. cloth, gilt head,$1 75
" ETHICS OF THE DUST. 12mo, extra cloth, gilt head, 1 75
" CROWN OF WILD OLIVE. 12mo, extra cloth, gilt head, 1 50
" DEUCALION. Collected Studies on the Lapse of Waves and Life of Stones. Parts 1 and 2. 12mo, cloth,...$1 00
" MORNINGS IN FLORENCE. Being Simple Studies on Christian Art for English Travellers. Santa Croce—The Golden Gate—Before the Soldan—The Vaulted Roof—The Strait Gate. 12mo, cloth..........................$1 00
" PROSERPINO. Studies of Wayside Flowers, while the air was yet pure. Among the Alps, and in the Scotland and England which my father knew. Parts 1 and 2. 12mo, cloth..$1 00

July, 1876.

IMPORTATION OF BOOKS, Etc.

AGENCY FOR THE SUPPLY OF

AMERICAN, ENGLISH, FRENCH, & GERMAN BOOKS,

PERIODICALS, &c., &c.

THE Subscribers continue to Import and to supply promptly and on the most favorable terms AMERICAN, ENGLISH, FRENCH and GERMAN BOOKS and PERIODICALS, in every department: MISCELLANEOUS, RELIGIOUS, and SCIENTIFIC.

They have constant communication with the principal American Publishers and Booksellers in the United States—have special agents in London and Paris, and direct correspondence with English, French, and German Publishers. Orders for a single Book or Periodical, or for Books and Periodicals in quantity, will receive their most careful attention.

ORDERS FOR FOREIGN BOOKS, &c.,

are forwarded as often as once a week, and answers may be looked for within six weeks. Catalogues and Bibliographical Works are kept for reference, and may be consulted at all times. Catalogues and Cheap Lists of particular Publishers are supplied gratis on application.

SPECIAL ATTENTION given to the procurement of RARE AND VALUABLE BOOKS, ENGRAVINGS, &c., for *Public and Private Libraries.*

BOOKS bound to order in ENGLAND and FRANCE by noted BINDERS for AMATEUR COLLECTORS.

BOOKS AND PERIODICALS can be mailed direct to any person or Public Library, from England and France.

BOOKS which have been published TWENTY YEARS may be imported free of duty.

PUBLIC LIBRARIES, SCHOOLS, AND COLLEGES, can import through us *two* copies of any Book, &c., free of duty.

OUR CHARGES FOR IMPORTING BOOKS ARE:

Per Sterling Shilling.................35 cents Currency.
Ditto, when free of duty.............30   "   "
Per Franc............................30   "   Gold.
Ditto, when free of duty.............26   "   "
Per Reichsmark.......................36   "   "
Ditto, when free of duty.............30   "   "

WHEN FROM SECOND-HAND ENGLISH CATALOGUES:

Per Sterling Shilling.................36 cents Gold.
Ditto, when free of duty.............30   "   "

JOHN WILEY & SONS,

PUBLISHERS AND IMPORTERS,

15 Astor Place, New York.

www.ingramcontent.com/pod-product-compliance
Lightning Source LLC
Chambersburg PA
CBHW032124230426
43672CB00009B/1850